COSMIC MYSTERIES

At the heart of Puppis A, the huge gaseous remnant of a supernova that exploded some 4,000 years ago *(radio map, inset)*, lies a small, mysterious structure *(red box):* three overlapping swirls of nebulosity, each dominated by a different chemical element and all expanding at a rate that suggests they are only 800 years old. The filaments—shown in the optical image at right, each element represented by a different color—might be the remains of a second, more recent stellar explosion within the shell of Puppis A. However, given the usual time scales of stellar evolution, astronomers are hard-pressed to explain how two neighbors could go supernova at almost the same moment.

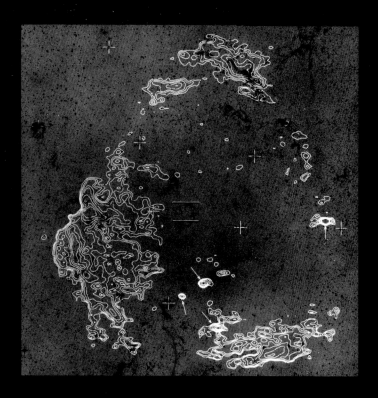

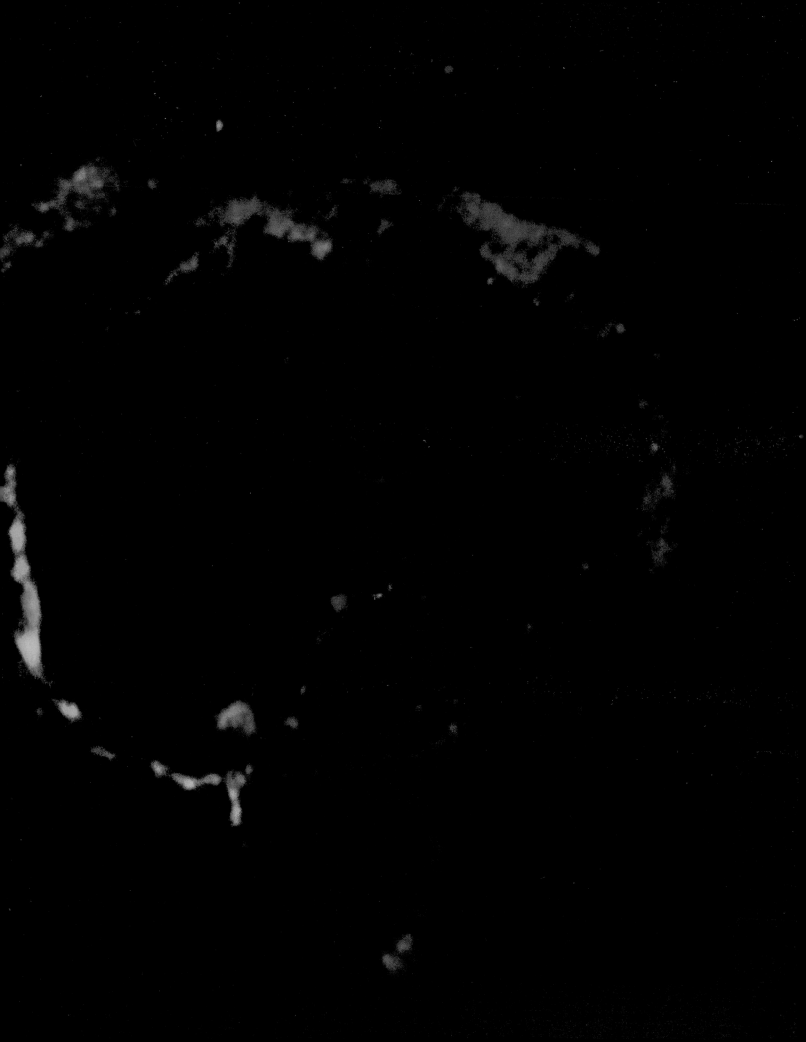

In the Southern Hemisphere constellation Centaurus lies a bizarre object called the Southern Crab for its uncanny resemblance to the creature. One tentative theory is that the body of the Crab may hide two stars: a hot white dwarf and a cool red giant. As the red giant swelled, emitting a dense wind of gas and dust that normally would have formed a bubble known as a planetary nebula, the orbital dynamics of the pair swirled the material into a vast disk around both stars and then into back-to-back streams that expanded into two bubbles. The glowing edges of the bubbles form the Crab's leglike filaments.

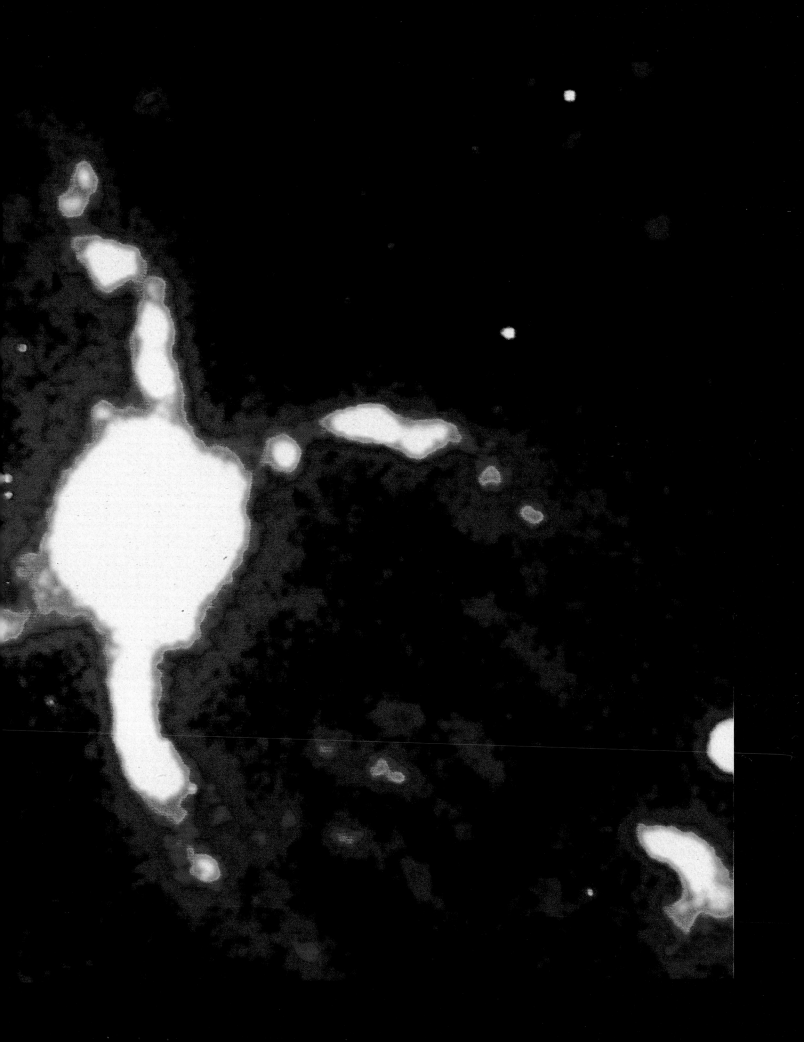

A color-enhanced optical photograph of the galaxy M82 in the constellation Ursa Major appears to justify the long practice of referring to it as the exploding galaxy. Now a reinterpretation of the evidence suggests that the turmoil in the galaxy's nucleus is not the result of an explosion but instead may have been caused by a close encounter—perhaps 40 million years ago—with M81, an enormous spiral galaxy ten times more massive than M82. The gravitational effects of M81's passage would have altered the orbits of millions of stars in the smaller system and jolted vast quantities of interstellar material out of its central plane, setting off a cycle of supernova explosions and a burst of star formation.

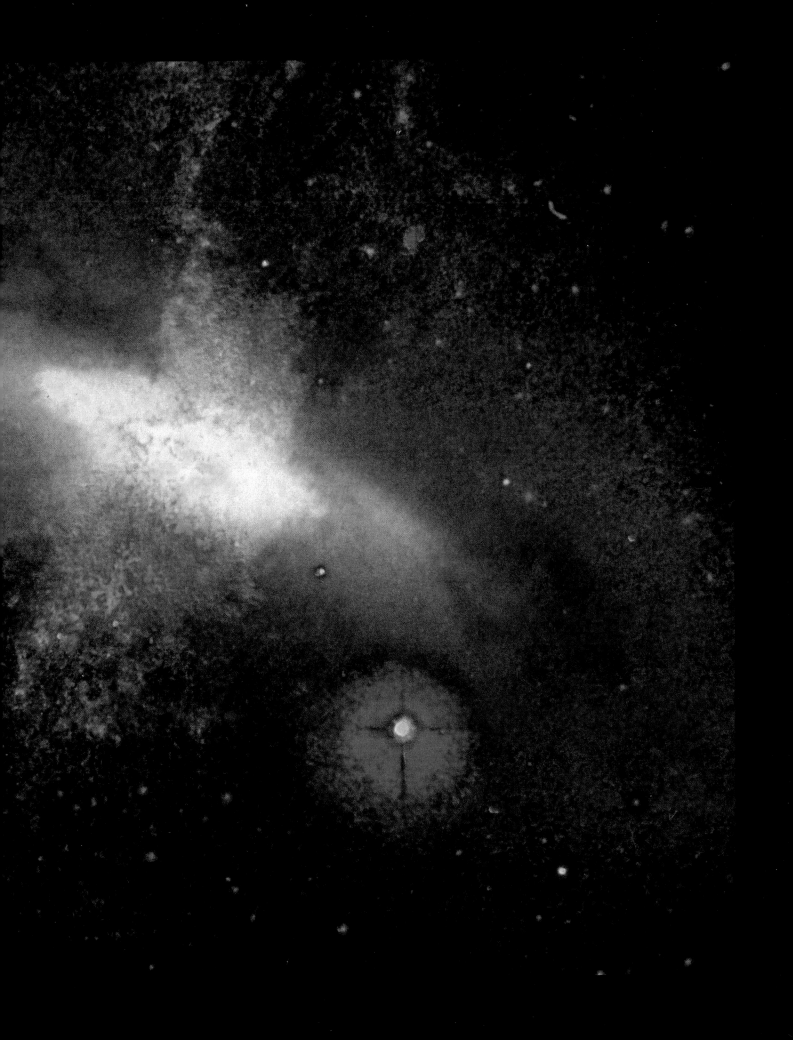

Other Publications:
THE NEW FACE OF WAR
HOW THINGS WORK
WINGS OF WAR
CREATIVE EVERYDAY COOKING
COLLECTOR'S LIBRARY OF THE UNKNOWN
CLASSICS OF WORLD WAR II
TIME-LIFE LIBRARY OF CURIOUS AND UNUSUAL FACTS
AMERICAN COUNTRY
THE THIRD REICH
THE TIME-LIFE GARDENER'S GUIDE
MYSTERIES OF THE UNKNOWN
TIME FRAME
FIX IT YOURSELF
FITNESS, HEALTH & NUTRITION
SUCCESSFUL PARENTING
HEALTHY HOME COOKING
UNDERSTANDING COMPUTERS
LIBRARY OF NATIONS
THE ENCHANTED WORLD
THE KODAK LIBRARY OF CREATIVE PHOTOGRAPHY
GREAT MEALS IN MINUTES
THE CIVIL WAR
PLANET EARTH
COLLECTOR'S LIBRARY OF THE CIVIL WAR
THE EPIC OF FLIGHT
THE GOOD COOK
WORLD WAR II
HOME REPAIR AND IMPROVEMENT
THE OLD WEST

This volume is one of a series that examines the universe in all its aspects, from its beginnings in the Big Bang to the promise of space exploration.

VOYAGE THROUGH THE UNIVERSE

COSMIC MYSTERIES

BY THE EDITORS OF TIME-LIFE BOOKS
ALEXANDRIA, VIRGINIA

CONTENTS

1 — 12 A Skyful of Puzzles
41 UNRAVELING STELLAR SIGNALS

2 — 52 Galactic Conundrums
84 MYSTERIOUS MOTIONS

3 — 94 Into the Unknown
122 LEAPS ACROSS SPACE-TIME

132 GLOSSARY

135 BIBLIOGRAPHY

139 INDEX

143 ACKNOWLEDGMENTS

143 PICTURE CREDITS

1/A Skyful of Puzzles

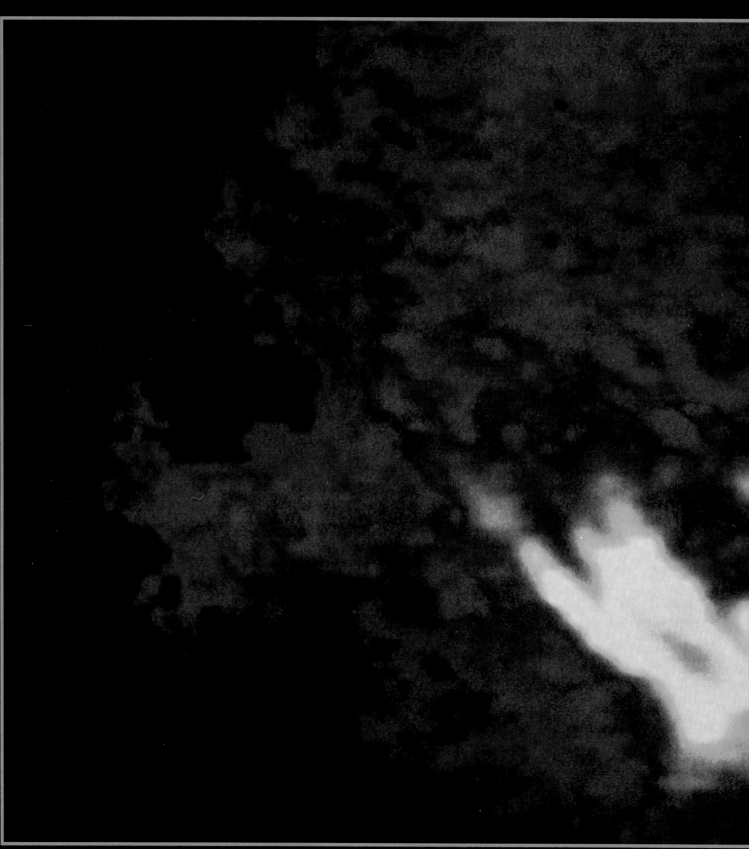

Conundrum at the core. Lying 30,000 light-years away, toward the constellation Sagittarius, the Milky Way's galactic center is blocked from optical view by interstellar dust and gas. But a radio image of the center reveals phenomenal activity: Gas near the galactic core is rotating at speeds approaching 54,000 miles per minute *(red)*—so fast that only an object on the order of five million solar masses could prevent it from simply flying off into interstellar space.

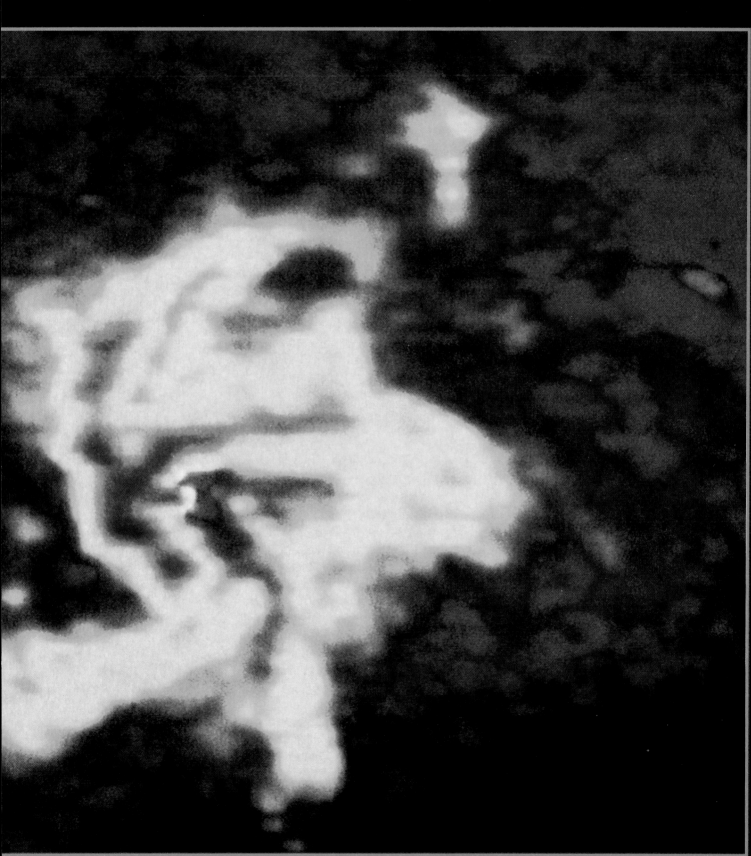

Surging through the galaxy at the speed of light, a wave of energy passed over the Solar System on March 5, 1979. No one on Earth noticed, but instruments aboard a variety of satellites and space probes scattered throughout the inner Solar System registered the awesome burst of electromagnetic radiation at a few minutes before 4:00 p.m., Greenwich mean time. Several hours later, flight controllers at Goddard Space Flight Center in Greenbelt, Maryland, spotted the pulse in a data transmission from the International Sun-Earth Explorer 3 *(ISEE-3)*, a solar research spacecraft stationed nearly a million miles sunward of Earth. They immediately alerted one of the mission scientists, Upendra Desai of Goddard's Laboratory for High Energy Astrophysics, who lived only two blocks from the control center. Desai was on the scene within minutes, but he need not have hurried. To this day, neither he nor any other physicist has put forth a satisfactory explanation of what happened.

Practically all that can be said for certain is that the event of March 5 was a pulse of gamma rays, the most energetic form of electromagnetic radiation, with wavelengths shorter than the radius of an atom. Gamma rays, which are undetectable from the ground because they cannot penetrate Earth's atmosphere, course through the universe from many different sources, constituting part of the omnipresent background radiation of the cosmos. During the twelve years prior to the 1979 event, scientists found that gamma rays also come in brief, high-intensity bursts that are many times more powerful than the background level.

The discovery was an accidental result of the United States' efforts to monitor the Soviet Union's compliance with the atmospheric nuclear test ban treaty of 1963. Since short bursts of gamma rays are an unmistakable characteristic of nuclear explosions, the U.S. Department of Defense launched four Vela satellites equipped with instruments designed to pick up gamma radiation. But when the Velas first began to register gamma ray pulses in 1967, the data triggered more confusion than consternation: The signals did not match those that had been produced by previous nuclear tests, nor did they seem to be coming from anywhere near Earth. The findings were so unexpected that physicists did not even report them in the scientific journals until 1973, when a group led by Ray Klebesadel of the Los Alamos Scientific Laboratory in New Mexico (renamed the Los Alamos National Laboratory in 1981) finally announced the existence of gamma ray bursts "of cosmic origin."

Although no specific sources could be identified, theorists believed they could explain what the sources were. When a dying star explodes as a supernova, its core collapses to form a rapidly spinning object composed almost entirely of neutrons packed together so densely that a teaspoonful of the material would weigh a billion tons. If less dense matter, perhaps from a companion star, were to fall toward the surface of such a neutron star, it would collect, concentrate, and then fuse under the tremendous pressure, producing a thermonuclear explosion that would release a brief but potent surge of gamma radiation. Between 1967 and the late 1970s, more than eighty separate bursts were detected, all of which fell within a range of intensities consistent with this sort of process.

The event of March 5, 1979, however, broke the mold in resounding fashion. To begin with, the recorded pulse was at least ten times more powerful than the largest previous burst. It reached maximum strength in less than a millisecond—hundreds of times faster than normal—and exhibited a decidedly unfamiliar decay pattern of rhythmic pulses at eight-second intervals, entirely unlike the irregular static of most fading bursts. But its most surprising feature was its apparent source. The standard theory of burst generation rested on the assumption that bursts originate somewhere within the Milky Way, Earth's home galaxy. To pinpoint the March 5 burst, an international team of physicists at Goddard and Los Alamos, in collaboration with a Franco-Soviet team, used a form of triangulation based on slight differences in the burst's arrival time at the nine spacecraft that detected it. They determined that the point of origin was a supernova remnant known as N49, about 170,000 light-years away in the Large Magellanic Cloud, a small companion galaxy of the Milky Way. On the one hand, the conclusion made sense, since the supernova remnant might well harbor a neutron star. On the other hand, if this huge pulse of radiation did come from N49, then the event that produced it must have been at least 100,000 times more powerful than the highest previous estimates.

The thermonuclear explosion hypothesis simply could not account for such violence. Loath to abandon an explanation that worked so well in every other instance, some scientists insisted that the real source of this particular burst had to be some unseen, much closer neutron star within the Milky Way that by chance fell directly in line with N49. Admittedly, the triangulation methods, while indicating direction precisely, did not reveal distance and thus left open this possibility. But the challenge was essentially a desperate one; given the overwhelming odds against such a coincidence in direction, the team of discoverers saw no reason to doubt that N49 was the source. As far as they were concerned, in the battle between theory and observation, it was theory that would have to give.

Thomas Cline, one of Desai's colleagues at Goddard, was foremost in promoting the view that an entirely new phenomenon had been detected, its nature still a complete mystery. As he noted in 1980, "very detailed and novel theoretical modeling" would be necessary in order to explain what had actually happened out there. A decade later, the mystery deepened. Between April and October 1991, a new satellite called the Compton Gamma Ray Observatory recorded 117 bursts of relatively uniform intensity, distributed evenly through space. Although scientists have proposed a number of theories to explain this distribution, they are still far from deciphering the bursts' bewildering cosmic message.

Frustrating as these events seem, they are the fuel that drives scientific inquiry. "When something ceases to be mysterious," writes theoretical physicist Freeman Dyson of Princeton's Institute for Advanced Study, "it ceases to be of absorbing concern to scientists. Almost all the things scientists think and dream about are mysterious." Often, strange objects or signals picked up in routine scannings of the sky represent no more than slight variations on well-understood phenomena, necessitating only minor tinkering with theory. At other times, observations are so baffling that theorists must start virtually from scratch to develop suitable explanations. Intriguing conundrums also originate on the theorist's blackboard. Spinning entire universes out of physics equations and quantum mechanical proofs, the deep thinkers of celestial science continually send observers scrambling to their telescopes and laboratories in search of concrete evidence to confirm or refute the latest theoretical constructs.

Still, tidy solutions are never guaranteed and specific issues tend to divide scientists into opposing camps. The disagreements sometimes become heated and may even endure for decades before new findings change the dynamics of the debate or, occasionally, settle the matter once and for all. Much of the time, on scales grand and small, the universe has simply refused to reveal its secrets, leaving theories unproved and observations unexplained.

The mysteries confronting astronomers in the late twentieth century range from subtle details of nearby phenomena to cosmic riddles in the deepest reaches of time and space. The familiar neighborhood of the Solar System remains a jigsaw puzzle with a number of missing pieces, and there are many strange doings still to explain among the Milky Way's panoply of stars. Farther afield, whole galaxies and collections of galaxies behave in ways that conflict with the accepted models for the evolution of the universe, and a host of exotic items and mind-bending possibilities—from quasars and black holes to time travel and multiple universes—lure scientists onward in the search for answers.

DEMISE OF THE CLOCKWORK UNIVERSE
Until recently, it had been an article of faith among scientists since Newton and Kepler that the universe was an orderly, predictable place, where the motions of every planet and star could be calculated with some degree of precision. But this deterministic, "clockwork" view of the cosmos has been challenged by the discoveries of twentieth-century science. Early on, the theories of relativity and quantum physics undermined the basic notion of

The Plunge into Chaos

For centuries, scientists were sure that forecasting the behavior of the weather, ocean currents, and other fickle-seeming phenomena would someday be possible. But during the 1960s, mathematicians studying even the simplest such systems discovered that the unpredictability could not be analyzed away; it was fundamental. Thus was born a new branch of science known as chaos theory—the study of randomness generated by fixed mathematical rules. In so-called chaotic systems, a minuscule change in one starting condition—the angle of a hand tossing dice, say—triggers enormous changes in outcome. Predicting the dice's final position requires perfectly measuring all initial conditions and re-solving the equations of motion with each tumble. In practice, no calculation or measurement can ever be precise enough.

Scientists today are finding evidence of chaotic behavior as far afield as Pluto's orbit. Pluto's gravitational relationship with Neptune introduces repeated tiny alterations in the planet's orbital equation that make it impossible to foretell its whereabouts 20 million years from now *(pages 20-21)*.

Yet chaos is not disorderly. Sometimes, out of turbulence, underlying order spontaneously emerges: A wave rolls unchanged over thousands of miles of ocean, or, in what are perhaps nature's most spectacular shows of order in chaos, great storms such as those on Jupiter, Saturn, and Neptune whirl undisturbed through seething planetary atmospheres—a mystery that scientists are only just beginning to unravel.

A study in stable chaos, Jupiter's Great Red Spot rolls like a planet-size ball bearing between opposing 200-mile-an-hour wind streams to its north and south, shown at left in a *Voyager 1* photo. Even though it inhabits this turbulent zone and is itself a drifting maelstrom of hurricane-like winds, the spot has persisted as a self-preserving entity for centuries.

Stability in the Midst of Tumult

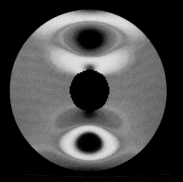
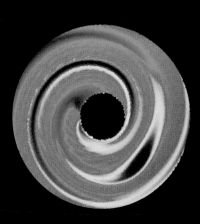
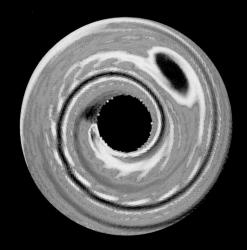
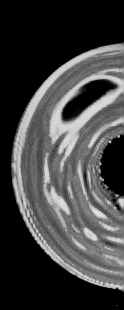

Freeze frames from chaos specialist Philip Marcus's Cray-computer simulation of Jupiter's Great Red Spot trace the evolution of order from chaos. As viewed from the giant planet's south pole *(top, black)*, a counterclockwise shear zone *(green)* forms between howling westward winds *(outer rim)* and eastward winds *(inner rim)* modeled on equations of motion for the Jovian atmosphere. Clockwise *(blue)* and counterclockwise *(yellow, red)* currents and vortices emerge out of the wind-driven turbulence; as wind velocity increases *(second frame)*, the blue vortex and yellow eddies, unable to withstand the 200-mile-per-hour gale, stretch into sinuous streaks. Sustained by the same chaotic energies that gobbled its blue companion, the red spot swirls tranquilly on. By the third and fourth frames, the blue wind streams fragment and disperse in the oppositely directed chaotic flow. The more robust yellow gusts organize into tiny whirlpools and are subsumed by the imperturbable giant spot.

In 1664, when English scientist Robert Hooke trained his telescope on Jupiter and spotted an enormous ellipse peering back like a sentient eye, he touched off a 300-year controversy. Was it a lava lake, a mountain plateau, or—more wondrous still—a monstrous helium egg, floating zeppelin-like through Jupiter's swirling clouds? Far-fetched speculations abounded until the 1979 visits of *Voyagers 1* and *2*, which revealed the spot to be a hurricane embedded in the planet's turbulent atmosphere. The mystery then shifted to its bizarre longevity: According to theory, the chaotic churning of the surrounding atmosphere should long ago have robbed the spot of its choreographed energy, rendering it as random as smoke in the wind.

Astronomer and applied mathematician Philip Marcus, now at the University of California at Berkeley, developed a mathematical model of the dynamics governing the red spot's durability, and then fed the data into a high-speed computer. Gradually, out of the programmed chaos on the computer screen, a robust, counterclockwise vortex emerged, preserving itself by siphoning energy from the ambient tumult. His results, shown here, were later verified by a team of researchers at the University of Texas at Austin using a mechanical device *(below, far right)* whose pumping and rotational action imitate the chaotic flow in the red spot's latitudinal region.

Both experiments produced remarkable likenesses of Jupiter's carmine eye. Their outcomes point to the same underlying dynamic: At a critical juncture, the feedback reverberating through a chaotic system can cause energy to bunch rather than to distribute equally among the turbulent elements. The energy becomes self-focused, limiting the patterns of interaction within the system. Over time, one chaotic fluctuation begins to dominate all others, just as Jupiter's Great Red Spot holds sway over the riotous zonal wind. Order emerges from chaos.

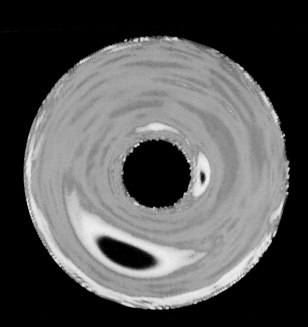

Maintaining the same latitude, periodicity, and relative size as its real-world counterpart, the red spot slips through the surrounding furor intact—proof that, given the proper initial conditions, a state of chaos can meld into stable, large-scale order.

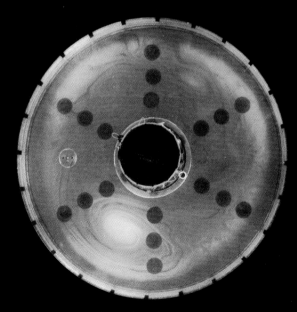

A laboratory experiment based on Marcus's computer model conjures the Great Red Spot. Brown dye injected into the three-foot, rapidly rotating, water-filled tank marks the helter-skelter motion of the simulated Jovian atmosphere but is unable to penetrate the spot's self-contained order.

In a strategy designed to expose the chaotic forces latent in Pluto's 20-billion-mile lap around the Sun, researchers plugged equations simulating near-identical orbits of two Plutos *(above)* into a special-purpose computer. After 20 million years, the orbital positions of the twin Plutos began to move apart. After several hundred million years had elapsed, the two planets were positioned on opposite sides of the Solar System *(right)*—a sure sign of chaotic processes at work.

Nothing so epitomizes order in nature as the planets' motions about the Sun. Yet, even within this clockworklike realm, scientists have detected intimations of chaos. Secreted in Pluto's orbit are the seeds of long-term unpredictability: Orbital analysts projecting only a few tens of millions of years ahead cannot pinpoint the planet's position. Their difficulty does not stem from any violation of Newtonian physics; rather, the planet's sensitivity to the conditions dictating its orbit at any moment is so extreme that even infinitesimal errors in measuring those conditions re-

Wisdom and Gerald Sussman, whose suspicions were aroused by Pluto's oddly tilted, elongated orbit *(below)* and its complex gravitational relationship with neighboring Neptune. To test their hunch, the researchers devised a special-purpose computer to evolve the orbits of Solar System bodies over a billion-year period. They then ran their orbital predictor twice for Pluto, choosing two minutely different starting points, four-millionths of an inch apart. If Pluto harbored a chaotic bent, they reasoned, the difference in starting positions would be mathematically amplified to gargan-

absolute measurements and introduced an unsettling element of chance into the description of such fundamental processes as the behavior of electrons. Then, in the 1980s, the new science of chaos hit the clockwork universe with the force of an earthquake.

"We are at the beginning of a major revolution. The whole way we see nature will be changed," maintains Georgia Institute of Technology physicist Joseph Ford, one of the first scientists to think about the origins of randomness in natural systems. Ford's early tinkering with the subject in the late 1950s inspired others to examine the issue in more detail; by the 1970s, the term chaos, supplied by mathematician James Yorke of the University of Maryland, was in widespread use among scientists even as the struggle to define it continued. According to the theory as it would eventually unfold, chaotic systems exhibit what can best be described as "deterministic randomness." The trajectory of a particle of cigarette smoke, for example, obeys simple equations yet is so complicated that it appears to be random: The particle's position in the future is unpredictable, even though it follows classic laws of motion. The significant aspect here is that the randomness is assumed to be intrinsic to the system and not the result of some accidental or extraneous effect. Scientists would have to learn to accept that the influences on some processes are so intricate that definitive predictions will always elude them.

Chaos theory, which was cropping up in disciplines as diverse as biology

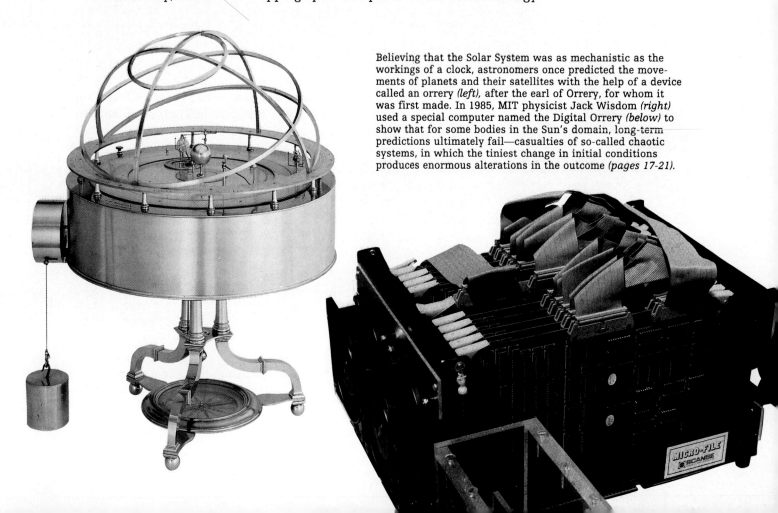

Believing that the Solar System was as mechanistic as the workings of a clock, astronomers once predicted the movements of planets and their satellites with the help of a device called an orrery *(left)*, after the earl of Orrery, for whom it was first made. In 1985, MIT physicist Jack Wisdom *(right)* used a special computer named the Digital Orrery *(below)* to show that for some bodies in the Sun's domain, long-term predictions ultimately fail—casualties of so-called chaotic systems, in which the tiniest change in initial conditions produces enormous alterations in the outcome *(pages 17-21)*.

and meteorology, was soon at work in astronomy as well. In the early 1980s, Jack Wisdom, a mathematician at the Massachusetts Institute of Technology, became interested in the trajectories of meteorites, some of which appeared to have arrived at Earth after being flung out of presumably stable orbits in the asteroid belt. Applying a new mathematical shortcut that greatly simplified the task of calculating complex interactions among multiple bodies, Wisdom discovered that the gravitational pull of Jupiter could disrupt the orbits of certain asteroids and send them plunging toward Earth. But Wisdom also found that the disruption occurred in a highly unpredictable fashion. The behavior could be traced to a basic principle of chaos theory known as a "sensitive dependence on initial conditions," which states that even the most minuscule change in initial conditions—such as an asteroid's orbital period in relation to Jupiter's—can lead to wildly different long-term results. No matter how precisely the orbit of one of these asteroids was calculated, there would always be uncertainty as to whether it would maintain its orbit or without warning go careering across the Solar System.

A DEDICATED CALCULATOR

Further studies uncovered chaos on a grander scale, in the orbit of one of the planets. Pluto attracted the attention of Wisdom and other scientists because its extremely elliptical orbit gives rise to the kind of intricate gravitational interactions—most notably with Neptune and Uranus—that can set the stage for chaotic behavior. This time employing a computer known as the Digital Orrery *(opposite)* to do the dirty work of calculating the orbital motions of the outer planets nearly a billion years into the future, Wisdom and colleague Gerald Sussman found that beyond about 20 million years, Pluto's orbit became unpredictable, a result once again of hair-trigger sensitivity to influencing factors. The orrery certainly generated a predicted position for Pluto every time the figures were run, but the slightest change in the initial conditions—smaller than could ever be detected in reality—would produce exponential changes in the results.

The implications bear on the Solar System's past as well as its future. Calculations backward in time indicate that Pluto's present orbit is of little help in ascertaining its historical behavior and that theories of its origin based on its current eccentricities may be impossible to prove. As for the ultimate effects of chaos on the Solar System as a whole, Wisdom and Sussman could draw but one conclusion: Since each of the planets has some gravitational influence on its siblings, and since Pluto's orbit is unpredictable, in the long run no planet's orbit is a sure thing.

Other examples of unusual orbital behavior among the denizens of the Solar System have led scientists to search for more tangible causes. Pluto itself was discovered by the American astronomer Clyde Tombaugh in 1930 as the result of a quest initiated because of observed irregularities in the orbits of Uranus and Neptune. Astronomers had believed for some time that the gravitational influence of an undetected planet was responsible for the perturbations, and with Tombaugh's find, the culprit seemed to have been apprehended. But by the late 1970s, improved estimates of Pluto's mass had revealed that it was far too small to disturb its giant neighbors, and scientists began to suspect that lurking somewhere beyond the ninth planet was a tenth—known, appropriately, as Planet X. A new generation of planet hunters took up the challenge. "I wish 'em luck," said octogenarian Tombaugh in 1988. "They're in for a lot of hard work."

Two of the leading workers are Robert Harrington and Kenneth Seidelmann, both of the U.S. Naval Observatory in Washington, D.C. Much of their energy has gone into figuring out just where to look, with computers playing a major role. Seidelmann's main contribution is to determine as accurately as possible the differences between the predicted and actual locations of Uranus and Neptune, known as residuals. He does so by comparing the latest sightings of the two planets with a computer simulation of their orbits based on observational data up to 1978. The measured discrepancies become the basis for more laborious rounds of modeling overseen by Harrington. Employing a program that simulates a ten-planet Solar System, Harrington plugs in different values for the mass, distance, and orbit of the unknown Planet X to see which will produce the observed effects on Uranus and Neptune. By the end of the 1980s, the computer had run through more than a third of a million simulations and come up with fewer than 200 possibilities.

The numbers seem to indicate that Planet X should currently be somewhere in a small patch of the southern sky near the constellation Centaurus, but searches of the area have so far turned up nothing. Some astronomers doubt they ever will. "Most of it's a bunch of nonsense," says Brian Marsden of the Harvard-Smithsonian Center for Astrophysics. "It is more likely that the residuals have some other explanation in some kind of observational error or instrumental error or systematic error. I think those have to be considered first before we jump to wild conclusions about distant planets." However, until any such errors can be brought to light, Planet X will remain a plausible albeit elusive quarry.

NEMESIS

At one time, a few theorists suggested that the Solar System might harbor yet another member, far beyond Planet X—and that it may be a killer. Fossil records show that about 65 million years ago, more than two-thirds of the species living on Earth, including all the dinosaurs, simply disappeared. In 1979, the late physicist Luis Alvarez and his son Walter, a geologist, proposed that the disaster occurred when a small asteroid or comet hit Earth, throwing a thick pall of dust into the upper atmosphere that cut off sunlight for several years; most plants would have died within months, disrupting the food chain and leading to the mass extinctions. Although the thoery created a storm of

The Case for Planet X

For more than half a century—and with varying degrees of intensity—some astronomers have proposed the existence of a tenth planet. Known as Planet X, the unseen world is invoked to explain certain anomalies in the behavior of Uranus, whose calculated orbit works for one circuit but not the next, and Neptune, whose predicted positions are noticeably in error after several years. Planet X, if it exists, might also explain the odd orbital dynamics of Neptune's moons Triton and Nereid, and the unexpected size and makeup of Pluto, which is so much smaller and rockier than its huge, gaseous neighbors.

The hunt for Planet X—based on calculations of its presumed mass and orbit—requires painstaking searches of designated patches of sky. In 1846 and again in 1930, such efforts paid off in the discoveries of Neptune and Pluto. Similar dedication may one day turn up a body that so far lives only in theory.

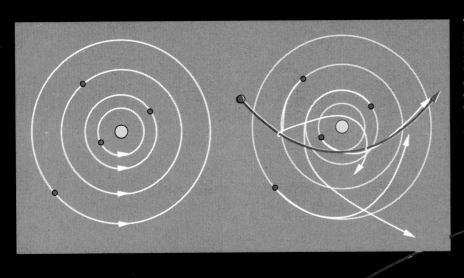

The fourth moon. One scenario to account for the dynamics of the outer Solar System hypothesizes that Neptune once had four major satellites—until Planet X *(red)* plowed through the system. The gravitational effects of its passage sent Triton into a backward circuit around Neptune, stretched Nereid's orbit out to six million miles, and knocked Pluto into an inclined and elliptical path around the Sun. Neptune's erstwhile fourth large moon was captured by the intruder.

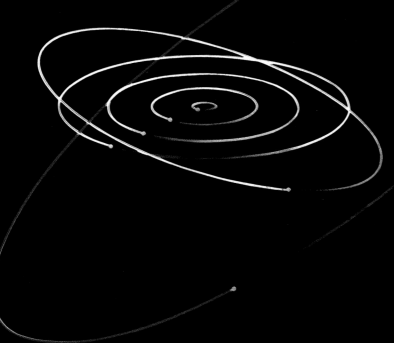

The first step. To find Planet X, scientists must first calculate its presumed mass and orbit. In the Solar System shown here, Planet X has a mass between two and five times that of Earth and travels in an orbit even more inclined and elliptical than Pluto's. At the farthest point in its orbit, Planet X would be 100 astronomical units from the Sun (or 100 times the distance from the Sun to Earth) and would take from 350 to 1,000 years to complete one circuit.

controversy when first introduced, it now seems well established—the most convincing evidence being that sedimentary layers from that epoch contain an unusual abundance of iridium, a metal that is otherwise rare on Earth—but is relatively common in meteorites.

But what caused the asteroid impact in the first place? Was it just a random event in the violent history of the Solar System, or was it part of a pattern? Through more recent, and more detailed, statistical analysis of the fossil record, geologists have found that mass extinctions seem to be periodic, occurring about every 26 million years. Such periodicity, suggested some, might be related to the passage of a massive object in a far-flung, extremely elliptical orbit around the Sun. Its gravitational pull, according to this theory, would regularly unleash a hail of potentially lethal comets and asteroids from the Oort cloud, a collection of frozen chunks of matter that astronomers believe encapsulates the Solar System.

The hypothetical body—known as Nemesis, or the Death Star—has drawn an even larger group of detractors than Planet X. For one thing, though predicted to be of stellar proportions, it would have to be very dim and cold to have escaped detection by both optical and infrared telescopes. And periodic extinctions have at least one other possible explanation: The entire Solar System is thought to oscillate up and down within the Milky Way, passing every 30 million years or so through the galactic plane, where clouds of interstellar dust and gas could send Oort cloud comets flying, or higher levels of cosmic radiation could directly affect Earth's climate. In any event, there is no other evidence for the existence of Nemesis beyond the extinction pattern, since Nemesis would currently be too far away to have any measurable effect on planetary behavior. So while Harrington and other Planet X hunters have been able to improve their chances with new data about the orbits of the outer planets, the dwindling numbers of Nemesis advocates can do nothing to narrow their search. The reasons for looking are too few and the places to look too numerous for Nemesis to have stirred much serious interest among members of the astronomical community.

INVISIBLE COMPANIONS

Despite the lack of hard evidence for a distant solar companion, the fact remains that solitary stars such as the Sun are relatively rare in the galaxy. About 80 percent of the stars visible in the Milky Way appear to be part of double, triple, or even quadruple star systems—most likely a consequence of the manner in which stars are born. Under certain conditions, such as the passage of a shock wave from a nearby supernova explosion, vast clouds of gas and dust in the spiral arms of galaxies may collapse inward upon themselves, creating regions of high density, pressure, and temperature. As the contraction proceeds, the cloud, or nebula, begins to fragment, and these dense regions form protostars—concentrations of mass whose gravity continues to pull in more dust and gas from the nebula's remnants. Eventually, pressure and temperature become high enough within these protostars to generate fusion reactions, and the newborn stars begin to shine. But before the nuclear furnaces get going, protostars themselves may fragment, the separate pieces typically evolving into stars that are almost certain to be locked in an orbital relationship throughout their history.

Of course, if the Sun does have a companion, it cannot be an identical twin or even an undersize sibling, whose light would be easily detected. One intriguing possibility is that Nemesis could be a strange and still-controversial celestial character known as a brown dwarf, an object that falls somewhere between planet and star. According to standard theories, planets form in a way that is different from the evolution of stars, accreting from small clumps of leftover dust and gas in a disk surrounding a protostar or newborn star. Their initial buildup results mostly from collisions with other bits and pieces in the disk, not through a collapse process or because of conditions of heightened pressure and temperature. Consequently, although they can grow quite large, they can never accumulate enough mass to initiate fusion reactions and become stars. But the theories also allow for another scenario: Some protostars, or perhaps protostar fragments, may never gather sufficient mass to ignite. Such a body would begin its life as a star does, yet, like a planet, would emit little energy of its own. It would be, by its popular moniker, a brown dwarf.

Brown dwarfs, if they exist, would satisfy theorists in at least two regards. For starters, they would fill a gap in the observed sequence of sizes of celestial objects. The largest known planet, Jupiter, has just one-tenth of one percent of the mass of the Sun, a medium-size star. Calculations indicate that the smallest stars require at least eight percent of a solar mass, or eighty Jupiters, to sustain fusion, and that planets could feasibly be as much as ten times larger than Jupiter. Brown dwarfs would cover the ground in between. They

The missing mate. Another hypothetical body, put forward to explain recurring mass extinctions on Earth, is known as Nemesis. This dark stellar companion to the Sun is theorized to be no more than .08 solar masses, with an orbit *(red)* that takes it out to 170,000 astronomical units, or about three-fifths the distance to the nearest known stars. But roughly every 26 million years, Nemesis returns to the Sun's neighborhood, passing en route through the Oort cloud—a huge sphere of cometary material, left over from the birth of the Solar System—and sending a rain of comets *(arrow)* toward the inner planets. The resulting meteoritic bombardment might have had catastrophic effects on Earth's climate and life forms.

might also help solve the enduring cosmological riddle of the universe's missing mass. Measurements of the gravitational forces at work within galaxies suggest that they should contain as much as ten times more matter than can be accounted for by observation. Proposers of brown dwarfs have reason to believe they should be common in the universe. A census of the skies reveals that very bright, very massive stars are relatively rare, but as mass and luminosity decline, the population increases. By extension, bodies with the postulated characteristics of brown dwarfs ought to be as common as grains of sand on a beach. The total bulk of so many brown dwarfs might be enough to account for the missing mass.

  Most scientists doubt the existence of a brown dwarf as a companion to the Sun, but several teams of investigators are hunting for them in the vicinity of other stars. Because brown dwarfs give off little or no visible light, the searchers have had to come up with ingenious methods for detecting their presence. A dark body orbiting a visible star ought to betray itself by creating

In 1989, University of Rochester astrophysicist William Forrest *(below)* found nine objects in the constellation Taurus that matched the characteristics of bodies known as brown dwarfs. Bigger than a Jupiter-size planet, but not massive enough to have triggered nuclear fusion and thus become a star, brown dwarfs would radiate in the infrared, making them very difficult to detect from Earth. The two shown in the infrared image at left *(arrows)* appeared to be associated with the star LkCa4 *(large shape at top)*.

a slight wobble in the motion of its bright companion, which would be revealed by a periodic shift in the star's spectrum. A group of Canadian astronomers looking for such spectral shifts with equipment designed to detect extremely minute variations announced in 1987 that they had found evidence for the existence of "low-mass companions" to at least two nearby stars and perhaps as many as seven. Bruce Campbell, speaking for the team, cautioned that they had determined nothing more than the probable masses of the unseen objects, which were between one and ten times that of Jupiter. The news, while exciting for its suggestion that planetary systems may be quite common, was discouraging for brown dwarf aficionados: Bodies of such mass were too small, and objects on the scale of brown dwarfs, which would create larger spectral shifts, ought to have been easier to spot. Having failed to detect any bodies in the required size range, Campbell was ready to rule out brown dwarfs entirely.

HOPEFUL SIGNS
Others questioned such a categorical conclusion, and within months of the Canadians' announcement, two observers reported that they had discovered something with at least one of the characteristics of a brown dwarf. Searching for infrared emissions—that is, levels of heat too weak to cause visible radiation—from bodies orbiting small stars, Benjamin Zuckerman of UCLA and Eric Becklin of the University of Hawaii found a clear-cut signal near the star Giclas 29-38. "It sticks out like a sore thumb," said Zuckerman. The object's temperature closely matched predictions for the thermal output of a brown dwarf, but Zuckerman remained cautious. "We don't know that we've seen a brown dwarf," he noted. "I hope we have. But no matter what it is, we've definitely seen something interesting that we have to understand."

Observations made by several other groups of astronomers in 1988 and 1989 seemed to hint at still more brown dwarfs, but the findings were ambiguous. Many of the candidates, for example, fell very close to the borderline between brown dwarf and star in terms of their temperature. Although the objects under scrutiny were apparently too cool to be stars, only a minor adjustment to the theoretical minimum stellar temperature would suffice for them to be classified as low-mass stars. Many astronomers felt safer broadening the definition of a known type of object than admitting the possibility of an entirely new sort of celestial entity.

In the summer of 1989, a group led by William J. Forrest of the University of Rochester found what seemed at the time to be strong evidence in support of brown dwarfs. Like Zuckerman and Becklin, Forrest's team set out to detect brown dwarfs by their heat; the difference was that they would look for ones thought to be at their hottest. During the first million years of its existence, a body destined to become a brown dwarf should be relatively bright at infrared wavelengths, glowing with the heat generated by its formation. Forrest concentrated the search in a section of the sky populated by young stars, on the assumption that nascent brown dwarfs might be present as well.

A stellar donation? In theory, the more massive the star, the faster it evolves. But middle-aged Sirius A *(below)* has a companion, Sirius B, that is half as massive yet is already a white dwarf. Sirius B (small white spot just below Sirius A) may once have been the larger star; then, as it passed through the red giant phase, it transferred much of its mass to Sirius A.

Using a new infrared camera at the NASA Infrared Telescope Facility on Mauna Kea in Hawaii, he and his colleagues found nine objects that seemed to match the predicted characteristics of young brown dwarfs. But Forrest was cautious: "We need more confirming evidence." The caution was justified. Subsequent spectral work on the best of Forrest's candidates suggests that the objects may be distant background stars whose high infrared readings are an artifact of being viewed through an intervening dust cloud. However, the objects' motion across the sky indicates that they are relatively nearby, in which case the infrared readings would support their status as brown dwarf candidates.

The lingering uncertainty over brown dwarfs suggests problems with the standard hypothesis. Even if brown dwarfs are actually found, the theories will need revising to explain, for example, the apparent absence of brown dwarfs in the vicinity of visible stars. Similar challenges have helped to shape the theoretical models not only of how stars begin their lives but also of how they age and die.

The current dogma on stellar evolution grew out of studies conducted early in the century by Danish astronomer Ejnar Hertzsprung and American Henry Norris Russell. Working independently, both men devised schemes for classifying stars on the basis of the only direct evidence available: brightness, which varies with a star's size and distance, and color, which varies with its temperature. For the sake of size comparison, brightness values were adjusted to discount the effects of distance. The resulting picture, known as the Hertzsprung-Russell diagram, shows that about 90 percent of all stars fall within a so-called main sequence, in which size and temperature are directly proportional—the brightest, largest stars being the hottest (colored white and blue), and the dimmest, smallest ones the coolest (orange or reddish). The remaining 10 percent of stars fall into two main groups off the main sequence: the hot but dim white dwarfs, and the red giants and supergiants, cool stars that are nevertheless bright because of their huge size.

Early guesses at stellar evolution postulated that stars simply move down the main sequence as they consume their hydrogen fuel, becoming smaller and colder and finally burning out. But this did not explain those odd red giants and white dwarfs. Later theorists, who were aided by discoveries about the internal dynamics of stellar fusion, clarified the picture. A typical star, it now appeared, spends most of its life at one spot on the main sequence, a position determined by its initial mass. The more massive a star is to begin with, the more quickly it consumes its nuclear fuel. As their core fuel dwindles and the core shrinks, some stars ignite an outer core of hy-

A multiple mystery. In the constellation Aquarius, R Aquarii (the rightmost of the two black spots in this false-color image) brightens and fades over a cycle of 387 days. But every forty-four years, the star's overall brightness dwindles for about eight years. One theory is that the star is a binary: a red supergiant that is partly eclipsed as it transfers gas to a white dwarf companion. The other black spot in the photo is thought to be a knot of gas ejected at high speed by the binary system.

drogen, causing it to swell and turning the star into a giant or supergiant. The outer layers of the star and this hydrogen shell eventually dissipate into space, leaving behind a small but very dense and hot core—a white dwarf—that will slowly shrink and fade out as it radiates its remaining energy. This broad outline of stellar evolution, itself evolving over several decades of theorizing, seems well established today; as the saying goes, however, the devil is in the details.

THE DOG AND PUP SHOW
One of the most familiar stars is a case in point. The brightest star in the night sky is Sirius, the Dog Star, a piercing bluish white dot located below and to the left of the constellation Orion, as viewed from the Northern Hemisphere. But there is more to Sirius than meets the eye. A suspicious wobble in its motion across the heavens led German astronomer Friedrich Wilhelm Bessel in 1844 to deduce the presence of an unseen companion. Eighteen years later, the father-and-son team of Alvan and Alvan G. Clark, opticians from Cambridgeport, Massachusetts, finally spotted Sirius B—the Pup—while trying out a telescope equipped with a powerful new 18.5-inch lens they had recently finished polishing. Calculations of the gravitational tug each star was exerting on the other revealed that the Pup had a mass about half as great as that of Sirius A but was 10,000 times dimmer and had to be vastly smaller. The facts baffled astronomers for more than fifty years, until physicists worked out the details of stellar evolution and could explain the discrepancy between mass and brightness: Sirius B was a white dwarf.

Unfortunately, this revelation only deepened the mystery. By rights, the more massive of the two stars, Sirius A, ought to have been the one farther along in its evolutionary cycle, the one that had burned its fuel more rapidly and moved from main sequence star through red giant to white dwarf. But Sirius A's vital statistics—its luminosity, color, and mass—put it right in the heart of the main sequence, a middle-aged star. Strange as it seemed, the less massive companion was the oldster.

Theorists eventually came up with what seemed a satisfactory explanation. Rather than trying to tinker with the accepted view of stellar evolution, they surmised that Sirius B was once, in fact, the more massive star, but as it expanded into the red giant stage, gravitational interactions caused a huge transfer of mass from B to A. There was a problem with this scenario, however: The two stars were not close enough to each other for such an exchange to have been possible. Then, in 1968, Dutch astrophysicist Edward van den Heuvel delivered a crucial insight. Through an elegant series of calculations, he demonstrated that mass transfer could actually have driven the two stars farther apart, into their present orbital relationship.

Although the Dutchman's paperwork made a convincing case on its own, clinching evidence soon came from observational data. Stars reveal their elementary composition in patterns of dark lines that appear in a star's spectrum when light of particular wavelengths is absorbed by various chemical elements in the star's outer atmosphere. Analysis of the absorption lines in Sirius A's light indicated high percentages of certain heavy metals that were unusual for a star of its class and age but in agreement with the hypothesis of an infusion of matter from Sirius B. The transfer theory was now safe for the textbooks.

The exchange of matter between binary stars such as Sirius A and B has become a fairly common explanation for bizarre or erratic stellar behavior, but unequivocal answers are rare. The red giant R Aquarii offers a particularly confusing picture. To begin with, it belongs to the intriguing class of stars known as variables, which brighten and then fade in a predictable pattern. R Aquarii reaches peak brightness every 387 days, a cycle that suggests it is physically pulsating, its outer shell heating and expanding, then cooling and contracting, as the star's energy winds down. But once every forty-four years, for a period of eight years, something else happens to R Aquarii. During these episodes, its maximum brightness decreases, and its spectrum changes.

In 1922, before all the details of the pattern had emerged, astronomers identified a variation in the red giant's spectrum as resulting from the presence of a small blue star, indicating that this was a binary system. At about the same time, a photograph taken from the Lowell Observatory in Flagstaff, Arizona, revealed a large nebula surrounding R Aquarii. Subsequent images taken many years later allowed astronomers to calculate the rate at which the nebula was expanding, which in turn enabled them to determine that hot gas had been ejected from the system some 600 years ago. In addition, later radio

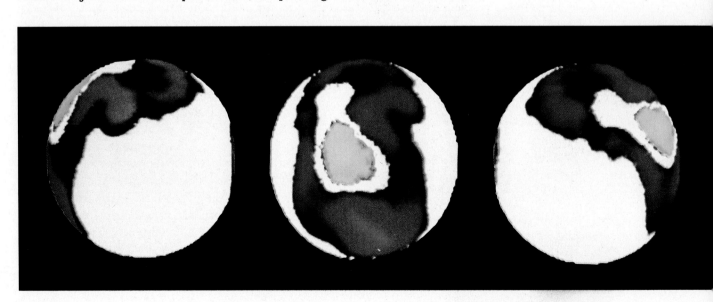

and ultraviolet detections revealed signs of smaller, more recent eruptions. Two theories have emerged to try to account for these phenomena. The first proposed that the blue companion star alone is enveloped in the large cloud, and together they orbit the red giant once every forty-four years, passing in front of it and partially eclipsing it for eight of those years, thus changing its spectrum and dimming it. This scenario depends on the eclipse's occurring when the two stars are farthest apart and thus moving at their slowest. An alternate theory predicts the exact opposite, that the eight-year period occurs when the binaries are closest together and results from a transfer of mass instead of an eclipse. With the gravitational tides at their strongest, gas from the swollen atmosphere of the red giant breaks away and is sucked in a spiral toward the companion, forming an accretion disk around it and causing eruptions where the cooler gas meets the companion's hot surface. The constant high-speed flow of material also makes the disk glow in a manner that would allow for the spectral changes, and the periodic loss of gas from the red giant would explain the diminished peaks in brightness.

CURIOUS CLOAKS

If only because it encompasses more of the system's oddities, many investigators lean toward the latter view of R Aquarii, but the galaxy is nonetheless rife with examples of binaries that are eclipsing each other. Almost every one of these systems has its own complex history, and general principles describing how such stars should interact often fail to account for specific situations. In the case of the Epsilon Aurigae system, theorists seem to be playing a constant game of catch-up with the facts. The primary star, Epsilon Aurigae A, is partially eclipsed once every twenty-seven years for about two years by a large cloud of material that probably harbors a small, dense companion. Since the eclipse is not total, astronomers initially proposed that the orbiting patch was very thin and flat, perhaps an accretion disk similar to the one suggested for R Aquarii.

But observations made during the star's 1982-1984 eclipse ruled out the thin-disk concept, and scientists began to suspect the presence of some unusual, semitransparent material expelled from one or the other of the stars. One candidate was plasma, a superheated gas that has been stripped of its electrons; however, the plasma should have been energized enough to radiate visible light, which was not observed. An alternative speculation—a loose swarm of solid dust particles, which would not completely blot out the light of the primary—also had problems. According to the laws of physics, such a disk ought to mimic the rings of Saturn, which are extremely thin. Yet the disk was estimated to be nearly 100 million miles thick.

In 1984, Cambridge University physicists Peter Eggleton and James Pringle came up with an attractive proposition: Rather than two stars, Epsilon Aurigae might actually be three, the large primary and two dense companions in close proximity to each other. In a binary system, the denser star would simply pull an accretion disk toward it, but with twin companions, complex

A SILICON PATTERN

In the computer-generated images at left, showing the surface of a star named Gamma Arietis at three different moments in its rotation, orange represents areas rich in ionized silicon, blue areas have none, and white indicates a normal amount. Scientists conjecture that the star's metal concentrations are linked to another, equally unusual, characteristic: The star rotates at nearly forty miles per second and has a strong, shifting magnetic field. But astronomers have yet to discover how the connection works.

gravitational effects would create a disk with a large vacant area around the denser pair, allowing some of the primary's light to peep through and be registered on Earth during an eclipse. And because the two smaller stars would keep the disk at a distance rather than collecting material from it, the disk could grow as massive as it seems to be. An additional twist—literally— was added in 1987 by Eggleton and Pringle's Cambridge colleague Sanjiv Kumar, when he suggested that the disk itself may be warped by gravitational forces, with one side tilted up like the brim of a hat and the other side tilted down. Such a configuration could account for the disk's apparent thickness. Until the two hypothetical companions are detected, however, the issue will remain unresolved.

FADE OUT, FADE IN

More curious still is the tale of Eta Carinae, a long-familiar inhabitant of the southern skies. Observers in the seventeenth century recorded Eta Carinae as a single, unremarkable fourth-magnitude star. (On a standardized scale, the dimmest stars visible without a telescope are sixth magnitude, very bright stars are first magnitude, and the brightest celestial objects—Sirius, Venus, the Moon, and the Sun—are given negative values.) But during the eighteenth century, it brightened to second magnitude, and by 1843 it had reached a magnitude of -1, outshining every other star in the sky except Sirius. Then it faded, until by the end of the century it was down to eighth magnitude, in effect winking out of sight. In recent years it has brightened again to about sixth magnitude, just barely visible to the naked eye.

During the 1830s, John Herschel, whose father, William, discovered Uranus, noted that Eta Carinae seemed to be surrounded by a nebulous cloud of gas. Because of its vague resemblance to a human form, later observers came to refer to the nebula as the Homunculus, or "little man," a term once used to describe a fetus; over time, the Homunculus has also changed, in shape as well as brightness. Detailed surveys in the 1960s and 1970s distinguished other features: an outer shell of gas beyond the Homunculus and a bright central core so thick that no single point of light can be discerned within it. The density of the core cloud and oddities in the spectrum of the region have called into doubt whether Eta Carinae is indeed a solitary star.

Some researchers think that Eta Carinae may be a stellar nursery, a region where young stars are forming, and that some of its strange variations may be a consequence of particularly violent births that would throw off huge clouds of obscuring dust and gas. This possibility was championed in 1976 by astronomer Paolo Maffei of the University of Perugia in Italy. "What seems fairly evident at this point," he wrote, "is that something is being born in Eta Carinae. It may be the kernel in which a multiple system of young stars is forming. It is even possible that the cocoon may contain a whole group of 100 to 200 stars."

But what was evident to Maffei was not at all apparent to other astronomers, many with their own theories about Eta Carinae. At various times,

Eclipse enigma. Every twenty-seven years, a mysterious object passes in front of Epsilon Aurigae A, a white supergiant 2,000 light-years away in the constellation Auriga, cutting its light by half for eighteen months. The illustration at right depicts one scenario for the nature of the object: a warped disk of particles around a binary pair that is in turn orbiting Epsilon Aurigae A. The disk, made up of gas sloughed off the supergiant, is warped by gravitational forces around the binary.

scientists have proposed that Eta Carinae is a highly evolved star in the process of ejecting its atmosphere; a very massive old star that, instead of becoming a white dwarf, is about to explode in the cataclysmic phenomenon known as a supernova; a star that has already been a supernova and is now in a postsupernova stage; or a middle-aged star that is rapidly stealing matter from an unseen companion. More recently, attention has focused on fast-moving clumps of gas and dust supposedly ejected from the outer shell. Nolan Walborn, a long-time Eta Carinae watcher at the Cerro Tololo Inter-American Observatory in Chile, has noted a similarity between the Eta Carinae clumps and those associated with Cassiopeia A, a suspected supernova remnant. The spectrographic identification of nitrogen in these clumps argues against

Maffei's star-birth interpretation, since nitrogen is rare in young stars but is manufactured in the interiors of evolved stars. So Eta Carinae, in this view, may well be a highly evolved star on the verge of going supernova. Given the awesome time scales of the cosmos, however, "on the verge" may mean as much as 10,000 years from now—a long time to wait for confirmation.

CLUES FROM A FATAL DANCE

The news is not always so perplexing or unsatisfying. Occasionally, the universe hands stymied theorists a tiny clue to some longstanding puzzle, such as the enigmatic behavior of pulsars. These objects release bursts of radio waves at such regular intervals that when they were first detected in 1967, some scientists thought they might be signals from extraterrestrials. Astrophysicists soon came up with more plausible explanations, among them the notion that pulsars were neutron stars, the dense, collapsed cores of stars that go supernova. According to a hypothesis first proposed by Thomas Gold of Cornell University, as a neutron star rotated, it would radiate energy in narrow beams, perhaps from its magnetic poles; if the magnetic axis happened to lie perpendicular to the axis of rotation and the alignment with Earth was just right, these beams would wash over Earth like the rays from a lighthouse beacon, creating the rhythmic pulsing. This view quickly achieved general acceptance, but in 1983, new evidence threw some doubt on the matter: Some of these objects, known as millisecond pulsars, were spinning far faster than they should, the beats separated by just a few thousandths of a second.

Basic physics held that while a neutron star was still contracting, it would pick up some speed—just as twirling ice skaters do by pulling in their arms. But contraction alone could not cause a body tens of miles in diameter to rotate nearly a thousand times each second. Theorists thus suggested that millisecond pulsars were acquiring angular, or rotational, momentum by pulling in matter from companion stars. Here again, however, the observational evidence was proving to be a problem: A substantial percentage of millisecond pulsars showed no sign of possessing a companion.

Then, in March 1988, Princeton University astronomer Andrew Fruchter and several colleagues made a crucial find with the world's largest radio telescope—the thousand-foot-wide dish at Arecibo, Puerto Rico. The dish itself, which is built into a natural hollow, is immobile and thus can view only a small patch of sky directly overhead, but by moving the receiving antenna suspended more than 400 feet above it, Fruchter was able to focus on one object in particular, a suspected pulsar that he had first spotted a year and a half before. Shortly after zeroing in on the target, Fruchter was surprised when the signal suddenly disappeared. He assumed an instrument malfunction, but technicians told him that all systems were working properly. Trying again a few days later, he had once more obtained the signal when, as he described it, "suddenly it just turned off. And then, about an hour later, it came back, and that's when we knew what we had." What

The case of the fading beauty. Once second in brightness only to Sirius, Eta Carinae, the star at the center of the Southern Hemisphere's Carina nebula, dimmed rapidly in the 1860s and then gradually returned to naked-eye visibility. An expanding shell of gas and dust, known as the Homunculus because of its vague resemblance to a human form (shown here in a color-coded optical image, with the star at the center), was probably responsible for blocking Eta Carinae's light. The puzzle is that the shell's composition is typical of ejecta from old stars, while the Carina nebula itself is a nursery for new stars. The paradox may be explained by Eta Carinae's size. At 120 times the mass of the Sun, the star may be nearing the end of a short, spectacular life that will climax in a supernova explosion in a few thousand years.

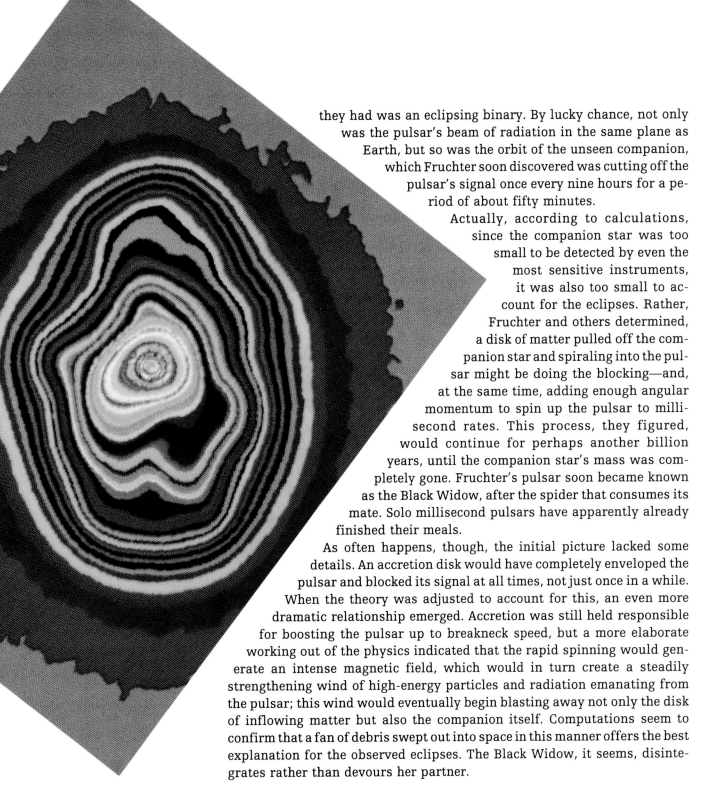

they had was an eclipsing binary. By lucky chance, not only was the pulsar's beam of radiation in the same plane as Earth, but so was the orbit of the unseen companion, which Fruchter soon discovered was cutting off the pulsar's signal once every nine hours for a period of about fifty minutes.

Actually, according to calculations, since the companion star was too small to be detected by even the most sensitive instruments, it was also too small to account for the eclipses. Rather, Fruchter and others determined, a disk of matter pulled off the companion star and spiraling into the pulsar might be doing the blocking—and, at the same time, adding enough angular momentum to spin up the pulsar to millisecond rates. This process, they figured, would continue for perhaps another billion years, until the companion star's mass was completely gone. Fruchter's pulsar soon became known as the Black Widow, after the spider that consumes its mate. Solo millisecond pulsars have apparently already finished their meals.

As often happens, though, the initial picture lacked some details. An accretion disk would have completely enveloped the pulsar and blocked its signal at all times, not just once in a while. When the theory was adjusted to account for this, an even more dramatic relationship emerged. Accretion was still held responsible for boosting the pulsar up to breakneck speed, but a more elaborate working out of the physics indicated that the rapid spinning would generate an intense magnetic field, which would in turn create a steadily strengthening wind of high-energy particles and radiation emanating from the pulsar; this wind would eventually begin blasting away not only the disk of inflowing matter but also the companion itself. Computations seem to confirm that a fan of debris swept out into space in this manner offers the best explanation for the observed eclipses. The Black Widow, it seems, disintegrates rather than devours her partner.

THE VERY YOUNG...
Winds from stars are becoming increasingly fascinating to astronomers, primarily because they appear to be intimately associated with both birth and death processes. Early in their lives, many stars of about the Sun's size may go through what is known as the T Tauri stage, named for a young star in the

constellation Taurus. As they make their last contractions not long after internal fusion gets under way, these stars are thought to unleash in all directions a steady, high-velocity flow of matter, strong enough in the case of the Sun, for example, to have stripped the inner planets of their primordial atmospheres. After this period of youthful exuberance, the flow calms down, most likely to the proportions of today's solar wind, whose streams of charged particles trigger the spectacular light show of the aurora borealis but leave Earth's atmosphere intact. Recent evidence, however, has given rise to a much more violent and much less orderly picture of star birth, and the T Tauri winds may be no more than gentle breezes in comparison to the hurricanes of energy that have been detected emanating from certain young stars. In the wake of the new findings, some scientists are beginning to wonder if the long-unchallenged theory of stellar evolution needs revising.

The first hint of something unusual came in the 1950s, when George H. Herbig, of the University of California at Santa Cruz, and Guillermo Haro, of the National Institute of Astrophysics, Optics and Electronics in Mexico, independently discovered several dimly glowing clumps of interstellar gas during routine optical surveys. Many other examples of these Herbig-Haro (HH) objects, as they came to be called, have since been found. Originally, they were thought to be protostellar condensations, or hydrogen clouds just beginning to form stars. Another interpretation saw them as simple reflection nebulae, patches of gas illuminated by the glow of nearby stars. But most of this early hypothesizing, hanging as it did on relatively uninformative optical images, was mere guesswork. By the middle of the 1980s, newly developed infrared instruments of unparalleled sensitivity were revealing a startling level of detail. Now HH objects appear to be clouds of molecular gas whose particles have been excited into radiance by the shock produced as they collide with the random molecules of gas and dust that fill all of interstellar space. They are moving at velocities approaching 250 miles per second, and their glow has been described as equivalent to the sonic boom produced by jet aircraft flying faster than the speed of sound.

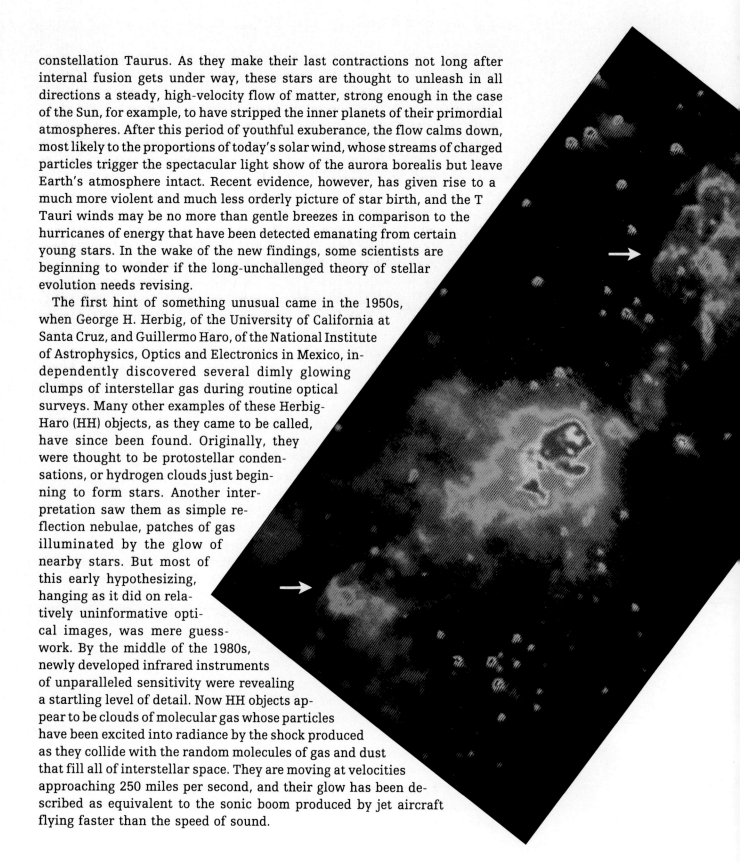

Most significant of all, bow-shaped HH objects were being found near very young stars, often on opposite sides of a star. All the evidence pointed to the inescapable conclusion that HH objects represent violent ejections of gas from newborn stars. These outflows were much stronger than T Tauri winds and also more discrete, typically forming two long, narrow, high-velocity jets aimed in opposite directions. They were also turning out to be quite a common phenomenon among new stars. The implications were clear. As astronomer Charles Lada of the University of Arizona put it, "A new stage of early stellar evolution of unanticipated astrophysical importance has been identified."

But theorists could do little more than describe the new data. It is not yet apparent how stars that are forming and therefore still gathering in surrounding matter could simultaneously be ejecting gas. Nor do physicists understand how HH objects are accelerated to such high speeds, or what mechanisms are responsible for focusing the outflowing matter into two narrow, oppositely directed streams. There is also some question as to whether the streams are continuous or broken by periodic ejections. Some HH objects are bullet shaped, as if they had been pumped from their stars in short bursts. In any case, says Lada, "very little is understood about the actual process of stellar birth. It is possible that the molecular outflow phase marks the end of star formation and the beginning of stellar evolution." If Lada is right, as astronomers learn more about these awesome outpourings, they may finally solve the enduring mystery of why stars come in such a wide variety of sizes.

Things that go bump in the night. On either side of a star-forming region called Cepheus A (colored red and orange in this infrared image), in the constellation Cepheus, are clumps of glowing gas known as Herbig-Haro objects *(arrows)*. These HH objects were named for George Herbig and Guillermo Haro, who independently discovered the phenomenon at optical wavelengths in the 1950s. Most are found in the vicinity of new stars, leading scientists to theorize that gas ejected by the stars during an early phase of development collides with atoms and molecules of interstellar gas, exciting them and causing them to emit radiation at specific infrared and optical wavelengths.

... AND THE VERY OLD

Investigations into the particulars of star formation inevitably lead to more general questions. Even as they struggle to understand how stars form in today's Milky Way, researchers want to learn more about how they may have formed at earlier stages in the life of the galaxy. Toward this end, astronomers have divided stars into three broad groups, or populations. Population I stars, scattered throughout the galaxy, are those that were born in the relatively recent past; they include the Sun, which is about 4.6 billion years old. Older stars, which came to life more than six billion years ago, are Population II stars; they are mostly situated in a spherical halo of star clusters around the outer reaches of the galaxy, as well as at the galactic center. Population III stars, the very first stars ever created, have yet to be found.

The key difference between the two known groups is that Population II stars show deficiencies in certain heavy elements, such as iron and calcium, a condition that holds clues not only to the makeup of the proposed Population III stars but also to the probable evolution of the universe. Cosmologists believe that most of the matter formed in the wake of the Big Bang—the explosion that brought all matter and energy into existence some 15 billion years ago—consisted of the two lightest elements, hydrogen and helium. Over the course of billions of years, after the first stars coalesced, heavier elements were fashioned by thermonuclear processes in the hearts of these primal bodies, or sometimes as a by-product of supernova explosions. The heavy debris from such explosions drifted through space until it was incorporated in the formation of new stars, and with each generation, the percentage of heavy elements increased. This explains why Population I stars, members of the youngest generations, contain the highest proportion of heavy elements.

Conversely, Population III stars should possess virtually none of the heavier elements found in stars like the Sun. Of course, having formed as little as half a billion years after the Big Bang and perhaps before the Milky Way took shape, they may be next to impossible to find. Massive Population III stars would long ago have exhausted their nuclear fuel. Only smaller, slower-burning stars with masses less than 80 percent that of the Sun could have survived into the present day. Searchers have found a few candidates in this size range that meet the statistical criteria, but many wonder if the standards are too loose; the examples spotted so far may come from early generations, but probably not the first.

The best place to continue the hunt may be that halo of stars at the very edge of the galaxy, where so many members of Population II reside. Because light takes time to travel from distant objects to Earth, the farther out one looks in space, the farther back one is actually looking in time, and thus the greater the odds that Population III stars will still be shining. Enticed by the possibilities of this built-in window on the past, some astronomers alter the focus entirely. Pursuing a more distant and more complex prey, they leave the whole galaxy behind for intergalactic space, an arena of infinitely greater proportions, where galaxies themselves are as numerous, as diverse, and often as unfathomable as the Milky Way's billions of stars.

UNRAVELING STELLAR SIGNALS

For most of human history, stars have represented all that is immutable in the heavens—faithful beacons in the constellations that make their annual pilgrimage across the sky. But astronomers have shown that stellar constancy is an illusion. Over periods of millions or billions of years, stars brighten and dim as they evolve. And some of these distant suns exhibit far more rapid fluctuations as well, flaring and fading in a matter of years, weeks, or days. Such changelings are known as variable stars.

The reasons for the brightness alterations of variable stars fall into two broad categories. Some stars seem to wink on and off because they are victims of occultation, eclipsed on a regular basis by unseen partners. In other cases, the changes are the result of intrinsic properties that are part of the star's dynamics. For example, irregular temperatures across the surface of a star might cause it to brighten and dim as it rotates.

Both mechanisms can be complex, but they are nonetheless susceptible to astronomical probing. As indicated on the following pages, the starting point of the detective work is a graph known as a light curve, which records the intensity of a star's brightness over time. From this clue, astronomers can construct surprisingly detailed theoretical models that might account for the stellar behavior. Still, a considerable amount of guesswork is involved, and scientists cannot always be sure that their answers to these cosmic puzzles have laid the matter to rest.

The Way of the Cepheids

One of the best-known variable stars is the yellow supergiant Delta Cephei in the constellation Cepheus. It brightens by a factor of two and then dims again in a regular rhythm lasting about five days. Scientists first noticed Delta Cephei's changeability in 1784; studies about 125 years later indicated that the changes correspond to fluctuations in the star's size. Delta Cephei swells and contracts by roughly 11 percent in a cycle identical in duration to the five-day brightness pattern. But the two rhythms are slightly out of phase: The star, shown in the composite image at left, is dimmest just after it reaches maximum size *(orange)* and brightest at a bit past minimum *(yellow)*.

Astronomers have deduced that the variations arise from fluctuations within a zone lying a few hundred thousand miles below the stellar surface, where helium acts as a valve. When the star contracts, the density of this layer increases, causing the helium to absorb radiation generated by thermonuclear processes in the core. The helium thus blocks the escape of radiant energy, and the star dims. But in time, heat energy builds up, stripping the helium of its electrons and decreasing the atmosphere's density. As the star expands, radiant energy is released, resulting in a dramatic brightening. With the continued escape of light, the pressure that caused the expansion begins to diminish, and the outer layers of the star's atmosphere succumb to the inward pull of gravity. As the star again begins to contract, its light suddenly dims.

Delta Cephei has given its name to an entire class of variables, the Cepheids. About 700 are known in the Milky Way, and many more have been detected in nearby galaxies. Because of a unique relationship between their intrinsic brightness and the period of their variability, Cepheids are invaluable for measuring cosmic distances. In the early 1900s, astronomers discovered that the longer a Cepheid's period of fluctuation is, the brighter the star. This link allowed calculation of the distance to Cepheids in the Great Nebula in Andromeda—proving for the first time that Andromeda is a galaxy separate from the Milky Way.

Contracting and expanding in a regular 5.34-day rhythm, the variable supergiant Delta Cephei brightens to a middling magnitude of 3.9 (about the same as the individual stars in the Pleiades) before dimming again to a magnitude of 5.1.

Microwave Beams from a Cloudy Shell

More than 1,600 light-years away in the constellation Pisces, an object called IRC + 10011 flares and fades over a period of about two years, but not in the visible range of the electromagnetic spectrum. Rather, IRC + 10011 emits most of its radiation at lower-frequency infrared and microwave wavelengths.

Scientists believe that the power source of IRC + 10011 is a red giant, a star that has consumed most of its hydrogen fuel and has become relatively unstable. Expanding and contracting in much the same way as a Cepheid, the star produces shock waves that push its atmosphere outward to a point where portions of the atmosphere condense to form a cloudy shell of dust and gas. The shell absorbs the star's visible radiation, reemitting much of it at infrared wavelengths. Radiation pressure eventually propels the obscuring material into space, where it disperses, but new clouds continue to form *(left)*.

More peculiar, however, is a kind of syncopated light curve *(below)* produced by microwave radiation apparently emitted by molecules of hydroxyl, or OH— coupled atoms of oxygen and hydrogen. The microwave radiation is believed to come from regions in the gas and dust shell where an excess of infrared radiation excites hydroxyl molecules in such a way that they are transformed into so-called masers, producing highly coherent, tightly focused beams of microwaves, all at the same frequency. (Maser stands for "microwave amplification by stimulated emission of radiation.") The microwaves' combined output is a powerful signal, detectable on Earth with radio telescopes.

Hydroxyl masers in a gas and dust shell around IRC + 10011 emit powerful microwave radiation in many directions, but radio telescopes on Earth can detect only those emitting along the line of sight. The double curve reflects the lag time between emissions of a pair of masers—one on the near side of the shell *(blue)*, the other on the far side *(purple)*.

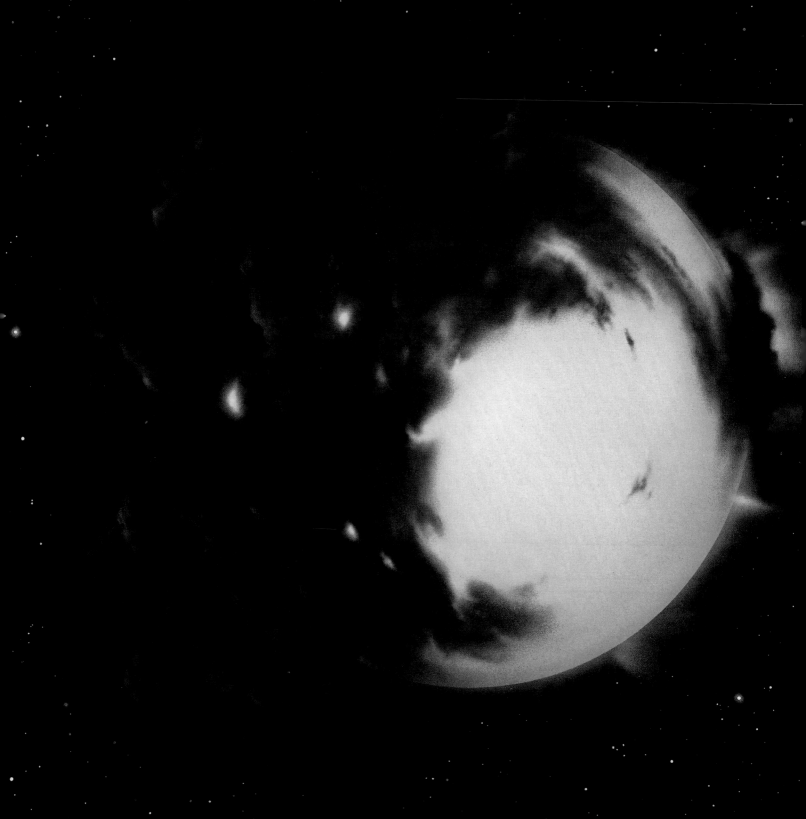

Exhalations of Stellar Soot

At unpredictable intervals, R Coronae Borealis, a yellow supergiant that is normally just barely visible without a telescope, undergoes a steep, sudden decline in brightness, sometimes dimming by a factor of more than 1,000 in just a few weeks. Then, in a process that may last for a few months or several years, it gradually regains its former shine. The recovery can be extremely erratic. During one episode recorded in the late 1800s, for example, the star took eleven years to return to maximum brightness, and its luminosity fluctuated wildly during the interim.

Although the exact reason for the precipitous dimming is not known, the star's spectral characteristics have led astronomers to postulate that clouds of carbon are the most likely cause. R Coronae Borealis and a few dozen similar stars, known as RCrB variables, are rich in helium and carbon, by-products of the thermonuclear reactions that have consumed much of their original hydrogen fuel.

According to the theoretical model illustrated here, hot, gaseous carbon at lower levels in the stellar atmosphere rises and cools, condensing into tiny particles of solid graphite. These grains gradually come together to form a thick cloud *(near left)* that blocks a portion of the star's light. Pushed outward by radiation pressure, the cloud spreads, obscuring more and more of the star's face *(far left)*. Eventually, however, the carbon cloud becomes patchy and increasingly transparent, allowing ever more light through—until the cycle begins again.

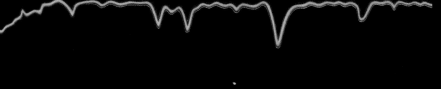

Over a ten-year span, the light curve of R Coronae Borealis fluctuates unpredictably, plunging abruptly by several magnitudes, then rising slowly with many intermediate dips in brightness. Some scientists have proposed that the star is playing peekaboo through thickening and thinning clouds of its own soot.

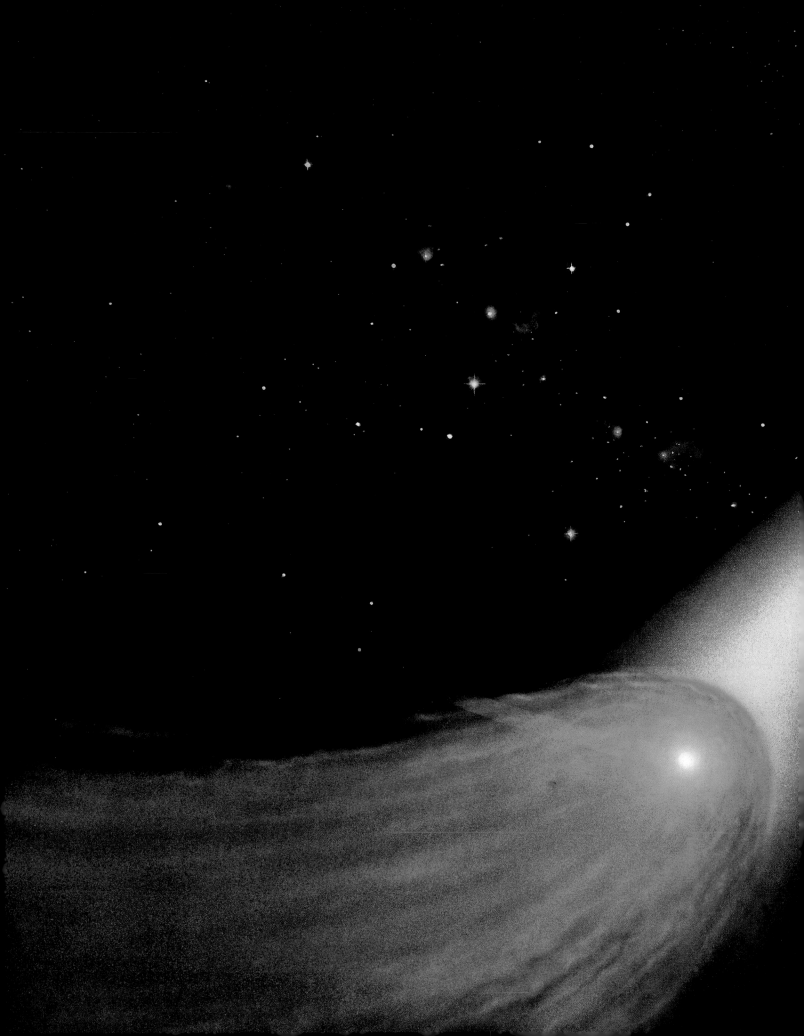

A Devouring Beacon

Some stars, rather than growing and diminishing in brightness, seem to switch abruptly on and off. First discovered in the late 1960s, these so-called pulsars emit regular, very rapid bursts of radio-wave energy—typically about once per second, but in the case of a millisecond pulsar, once every few thousandths of a second. While theorists were quick to develop an explanation for the slower-paced variety, they have been hard-pressed to account for the dizzying rate of millisecond pulsars.

Pulsars in general are believed to be neutron stars, the tiny, incredibly dense remnants of supernovae. When a dying star explodes into a supernova, the momentum of its original rotation is concentrated in the surviving inner core, which begins to spin rapidly, emitting radio waves along its magnetic axis. The rhythmic blinking of a pulsar occurs only if the magnetic axis lies perpendicular to the axis of rotation and is aligned so that the beams of radiation strike Earth. But what makes a millisecond pulsar spin so fast? The likeliest conjecture is that matter from a companion star, pulled by the neutron star's intense gravity, spirals down onto it, adding momentum and increasing the pulsar's rotation rate. Yet in some cases, millisecond pulsars have been found without a partner.

In March 1988, however, Princeton astronomers detected a millisecond pulsar with a curious hiccup in its rhythm that may hold the key to the puzzle. Every nine hours, the pulsar's rapid-fire beat disappears for fifty minutes. Because the interruption is clearly too long to be caused merely by the passage of an eclipsing companion, astronomers believe that a swath of matter evaporated from the surface of such a companion is blocking the signal. Blown off by the intense radiation from the pulsar, the vaporized star stuff forms a broad tail that spreads into interstellar space. After a few billion years, this so-called Black Widow pulsar will completely destroy its mate—a white dwarf so small and dim it at first escaped detection—and will then join the ranks of solo millisecond pulsars that were once inexplicable.

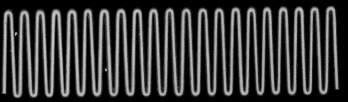

The Black Widow pulsar, a rapidly spinning neutron star, emits pulses of radio waves once every few milliseconds (frequency not to scale). But for fifty minutes every nine hours, its signals are eclipsed by the passage of a cloud of stellar matter blown off its white dwarf companion by the pulsar's own radiation.

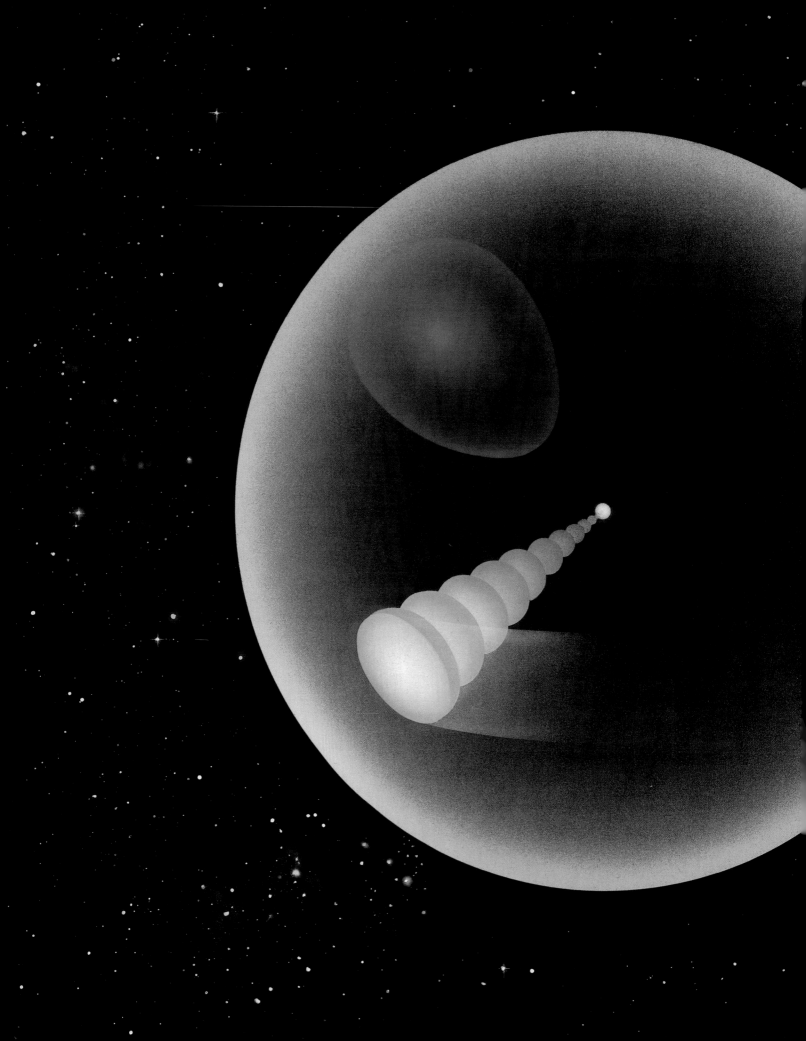

A Flashing Plasma Searchlight

Over the course of a year, a number of stars will blaze up like cosmic flashbulbs in the sky. These phenomena, known as novae, occur in close binary systems, where the greater gravity of one star draws matter from its less substantial companion. As the matter accretes onto the surface of the more massive star, temperature and pressure increase until finally they trigger nuclear fusion, which usually happens only in stellar cores. Incandescent matter is blasted into space, and the star brightens by a factor of many millions as a result.

One recent nova provided both a public show and an astronomical puzzle. Nova Cygni 1975 burst into view in the year of its name with such brilliance that it could be seen even in the night skies of light-drenched cities. Like all novae, it began to fade quickly. But the generally falling light curve was interrupted by an odd uptick every three hours and twenty minutes.

The mystery persisted for thirteen years, until a team of researchers discovered evidence—in the form of polarized light—that the white dwarf at the center of the nova possessed a magnetic field more than 10 million times stronger than Earth's. This magnetic field, they theorized, funneled the accreting matter to one of the dwarf's magnetic poles, thereby concentrating the buildup of pressure and temperature that would ultimately set off the nova blast on a small portion of the dwarf's surface. When the explosion occurred, the same magnetic forces channeled bubbles of the hottest and densest matter out from the polar region. Rotating with the star, the jet of plasma pulses produced a radiant spot on the envelope of matter blasted away from the star. Then, as the envelope contracted of its own gravity, the pole remained the most luminous region, flashing like a searchlight on each stellar revolution.

The light curve of Nova Cygni 1975 (not drawn to scale) shows the expected rapid fading from its peak of brightness—with one unique exception: Every three hours and twenty minutes, the star's light intensifies briefly before continuing its slow, downward plunge.

2/Galactic Conundrums

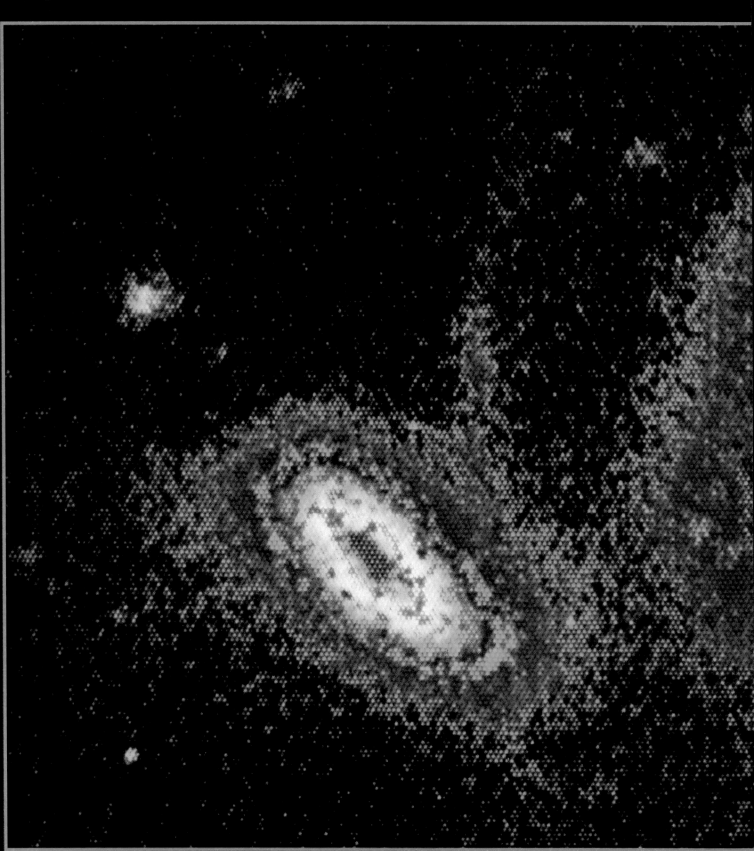

Clashing titans. Disturbed by their warring gravitational forces, spiral galaxy NGC 5754 and a small galaxy known as Arp 297C have initiated a burst of star formation (yellow in this false-color optical image)—or so astronomers surmise. Scientists try to deduce the evolution of such interacting star systems by catching a variety of them in midact, looking for such telltale signs as streamers of gas *(bluish arc)* that seem to be pulled into intergalactic bridges.

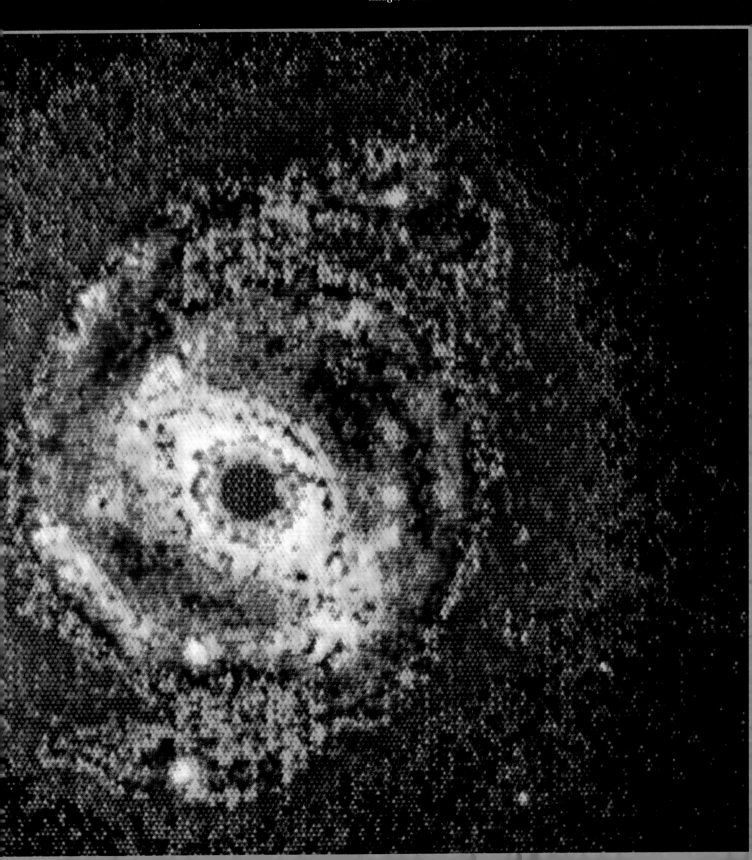

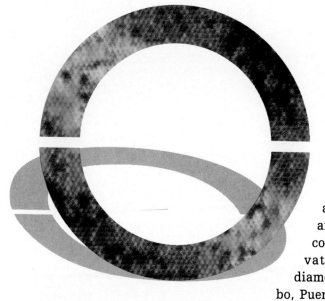

One night in the spring of 1989, astronomers Martha Haynes and Riccardo Giovanelli were conducting some routine observations with the 1,000-foot-diameter radio telescope at Arecibo, Puerto Rico, the same giant dish that had revealed the odd behavior of the Black Widow pulsar to Andrew Fruchter a year earlier. Following common practice, they decided in the middle of the session to recalibrate the detecting equipment against a quiet background—a region entirely devoid of radio-emitting objects. The most convenient spot at the time was a supposedly empty patch of sky just to the south of the constellation Virgo.

The telescope's receiver should have picked up absolutely nothing. Instead, the recording apparatus began to trace out the unmistakable radio signature of hydrogen, at a level of intensity typically associated with the presence of a galaxy. At first, neither Haynes nor Giovanelli got too excited. Like Fruchter before them, they suspected the peculiar signal was due to the sort of electronic gremlins that often plague earthbound radio astronomers. "We thought it was just radar interference from the airport," says Giovanelli, "or somebody's garage door opener."

But on subsequent nights, the same signal appeared in precisely the same location. Something was definitely out there. Over the next few months, the two astronomers compiled dozens of readings from all around the point of the original detection, mapping out the structure of an enormous cloud of hydrogen ten times larger than the Milky Way. It had the rough elliptical shape of a galaxy, with two dense clumps of gas near the center, but there was no evidence of any stars. By the end of July, Haynes and Giovanelli were ready to share their findings. Purely by chance, they had discovered an object that had eluded detection since its existence had been proposed more than fifteen years earlier: an embryonic galaxy on the verge of bursting into stellar life.

What made the so-called protogalaxy such a surprise was its closeness. The cloud was only 65 million light-years away, almost next door in astronomical terms. This challenged one of the fundamental assumptions about the creation of galaxies. Standard theory holds that galaxies materialized relatively early in the history of the universe, coalescing out of the cosmic soup of hot, thin gases and, no later than 10 billion years ago, becoming dense enough for

Nascent galaxies? A vast hydrogen cloud only 65 million light-years away could upset accepted theory that galaxies formed all at once about 15 billion years ago. On the radio contour map at right, two regions marked by heavy concentrations of atomic hydrogen *(darkest blue)* suggest that the cloud is a gaseous envelope around two dwarf galaxies in the making, a supposition bolstered by optical evidence for a tiny source of starlight corresponding to the denser of the two hydrogen clumps. The overwhelming preponderance of gas over stars—99 percent to 1 percent—is taken as further evidence that the cloud is a late-blooming protogalaxy.

stars to form. Thus, the only protogalaxies currently detectable should be a great distance away, so far that their radiation would take more than 10 billion years to reach Earth. One as nearby as Haynes and Giovanelli's should have begun to shine with stars long ago.

But another strange characteristic might account for the anomaly: Measurements indicated that the cloud contained so little mass—about a tenth as much as the Milky Way—that if it had been any less dense, it would not have had sufficient gravitational force to hold itself together, let alone contract. The two discoverers therefore concluded that their cloud had indeed formed at the same time as other protogalaxies, but because of its low mass, it was developing at a snail's pace and was only now approaching the density needed to produce stars. If this was true, not only would astronomers have learned valuable new information about the dynamics of galaxy formation, but for the first time they would be able to study matter that was essentially unchanged since the beginning of the universe, as if held in a kind of suspended animation. The response from the astronomical community was immediate and, for the most part, enthusiastic. To astronomer and protogalaxy specialist Arthur Wolfe, of the University of California at San Diego, the discovery was "like finding a mummy or a new pyramid, unaltered from the past."

Others reacted with more caution, and some with outright skepticism. In fact, within a week of the announcement, Mike Irwin and Richard McMahon of the University of Cambridge in England had come up with apparently refuting evidence. Scanning a series of photographic plates made during the 1970s, they found a handful of stars in the same reportedly starless region, right in the midst of the denser of the two clumps of hydrogen on the radio maps. To the two British astronomers, this suggested that Haynes and Giovanelli had seen not a protogalaxy but rather a large gas envelope surrounding a small, already formed irregular galaxy. "They jumped to a conclusion because they did not see an optical counterpart," said Irwin.

But the debate continued. As far as Giovanelli was concerned, Irwin and McMahon were no less guilty of rushing to judgment. Even if there were a few stars within the cloud, he countered, it could still qualify as a protogalaxy or, in the purest terms, a galaxy in its very earliest stages, and would still challenge the accepted timetable for galaxy formation. Yet so painstaking a science is astronomy that, even after months of intensive study, Haynes and Giovanelli were the first to admit that the matter remained unresolved.

CRUCIBLES OF CONTROVERSY

The opposing camps on the issue of protogalaxies have so far managed to keep their arguments relatively civil, a surprisingly rare feat in the field of galactic studies. More so than stars, galaxies tend to evoke almost primal responses among professional sky gazers. As the largest discrete objects in nature, galaxies represent the basic units of the universe. Incubators of stars, generators of energy, vast repositories of invisible matter (perhaps), and definers of large-scale structure, galaxies are central to almost every major theory about the nature of the cosmos. It is little wonder, then, that they have been at the heart of nearly every great controversy in astronomy throughout most of the twentieth century.

A large part of the problem is the inconceivable remoteness of galaxies, the nearest being more than 150,000 light-years away and the farthest many billions of light-years distant. As a result, they have been very slow to yield their secrets, often forcing scientists to develop hypotheses on the barest minimum of information and leaving plenty of room for argument. For hundreds of years, galaxies were a complete mystery to astronomers, inexplicable fuzzy patches among the stars' sharp points of light. Eventually, in the nineteenth century, more powerful telescopes resolved individual stars within these so-called nebulae, and for the first time theorists began to consider the possibility that there were such things as galaxies, separate collections of stars that existed beyond the bounds of the Milky Way. Yet the evidence was so sketchy and distances were so difficult to estimate that astronomers did not reach a consensus on the subject until the 1920s, with many reputations rising and falling in the interim as the weight of scientific opinion shifted from one view to another.

Technological advances throughout the twentieth century have revealed nearby galaxies in greater and greater detail, as well as enabling astronomers to peer ever deeper into the immensities of intergalactic space. Yet each new discovery, rather than resolving arguments, seems to raise new doubts and new debates. Not long after the existence of galaxies was widely accepted, for example, fresh arguments arose, with one side contending that there were just a few standard types of galaxies, and the other, that galaxies came in an infinite variety of shapes and sizes. Similarly, in the 1960s, fierce debates raged over the cosmic creatures known as quasars. Were they the powerful hearts of the most distant galaxies, or had nature conjured up a bizarre trick to tempt and tease unwary astronomers? More recently, a host of weird and wonderful galaxies—colliding and collapsing, exploding and imploding, even cannibalizing their partners—have led experts to new disagreements.

As the twentieth century draws to a close, the largest questions loom.

Increasingly sophisticated surveying techniques have revealed new and baffling architectures in the cosmos. Once thought of as isolated "island universes," galaxies are now known to form chains of islands, and the chains, in turn, to create huge starry archipelagos. Furthermore, conglomerations of galaxies seem to be stretched into thin sheets surrounding gigantic bubble-like voids that even the best minds in theoretical physics are hard-pressed to explain. Additional evidence points to the existence of unimaginable accumulations of galaxies with a combined gravitational pull strong enough to influence other galaxies hundreds of millions of light-years away. As so often before, these findings are raising passions and stirring intense debate, with no less at stake than an understanding of the origin, evolution, and eventual fate of the universe.

CHARTING THE ISLAND UNIVERSES
The final determination that there were other star systems besides the Milky Way owes as much to the workings of a great instrument as it does to the work of a great astronomer. The astronomer was Edwin Hubble, and the instrument, the 100-inch optical reflector telescope at the Mount Wilson Observatory in southern California. In 1919, Hubble, a talented young scholar who had passed up a promising career as a lawyer to study astronomy, joined the Mount Wilson staff after a two-year tour of duty with the American Expeditionary Force in France. Hubble had done his doctoral thesis at the University of Chicago on spiral nebulae, one of the most intriguing forms of the hazy patches of light that had perplexed scientists for so many years. Convinced that the spirals were indeed other galaxies, he realized that the recently completed giant reflector, then the largest in the world, would be the perfect tool for proving the point.

Because of its large diameter, the new telescope would not only collect enough light to pick out very faint objects but also be able to resolve them more clearly, distinguishing individual stars within nebulae better than ever before. Hubble set his sights on the promising Andromeda nebula, the only spiral nebula visible to the naked eye from the Northern Hemisphere, searching diligently for examples of the stars known as Cepheid variables *(pages 42-43)*. Earlier, Harvard's Henrietta Leavitt had shown that these peculiar stars could serve as a kind of astronomical yardstick, because they had the unique property of dimming and brightening at a rate directly linked to their intrinsic luminosity: The slower their rate of variation, the greater their maximum brightness. And if one knew how bright a star really was, then its apparent brightness as seen from Earth would provide an indication of its distance according to the inverse-square law, which states that an object's brightness diminishes in proportion to the square of its distance.

By good fortune, in 1923, Hubble found a Cepheid in Andromeda. Using Leavitt's period-luminosity law, he estimated that Andromeda was some 800,000 light-years away, or approximately eight times the distance to the farthest known star in the Milky Way. Clearly, this spiral nebula was not part of the Milky Way structure. (More accurate measurements later revised the distance to Andromeda to about two million light-years.)

In a single stroke, Hubble had revolutionized astronomy, broadening the very dimensions of space and demonstrating that galaxies were a fundamental constituent of the universe. Today, however, he is best remembered for an even more extraordinary revelation. Astronomers had known for some time that light from a moving object undergoes the same kind of wavelength shifting, or Doppler effect, that changes the pitch of a siren as a police car or ambulance speeds by: Light from objects approaching Earth is shifted toward the shorter-wavelength blue end of the spectrum, and light from receding objects, toward the longer-wavelength red end. The greater the shift, the swifter the rate at which an object is approaching or receding. In 1929, again using the 100-inch telescope, Hubble and his assistant Milton Humason recorded the spectra of scores of galaxies, including many whose distances had already been derived from Cepheid observations. Not only were the majority of them redshifted—something first discovered several years earlier by Vesto M. Slipher, a spectroscopist at the Lowell Observatory in Arizona—but in every case, the degree of redshift was directly proportional to the galaxy's distance from Earth, with more remote galaxies receding faster. By calculating the steady rate at which this recessional velocity changed with distance—a figure known ever since as the Hubble constant—Hubble established a means for measuring the distance to any galaxy in the universe, including the many that were too far away for Cepheids to be visible. (Ironically, precise measurements have remained problematic because astronomers have had to continually adjust the Hubble constant as they revise the baseline distance measurements from which it is calculated.)

Although astronomers relished this first-ever opportunity to take the measure of the visible universe, the news was also profoundly unsettling. If all galaxies were indeed flying apart, then the universe itself must be expanding in all directions, as if flung outward by some gigantic explosion. Cosmologists, it seemed, would have to start from scratch, abandoning the long-standing assumption that the universe was immutable and eternal, and incorporating into their theories the possibility that it had had a distinct beginning, had been changing ever since, and might one day come to an end.

TYPECASTING THE GALAXIES

So radical was the concept of an expanding universe that many astronomers initially refused to accept it. Hubble himself never seemed completely comfortable with the idea, which may account for the fact that he soon shifted his attention away from distance studies and spent the remaining thirty years of his life concentrating on the properties and peculiarities of individual galaxies. Classifying hundreds of bright galaxies that had been photographed with the trusty 100-inch at Mount Wilson, Hubble divided his specimens into two main categories: spirals and ellipticals. Nearly two-thirds were typed as the former—great bluish white pinwheels, with bright central bulges of close-

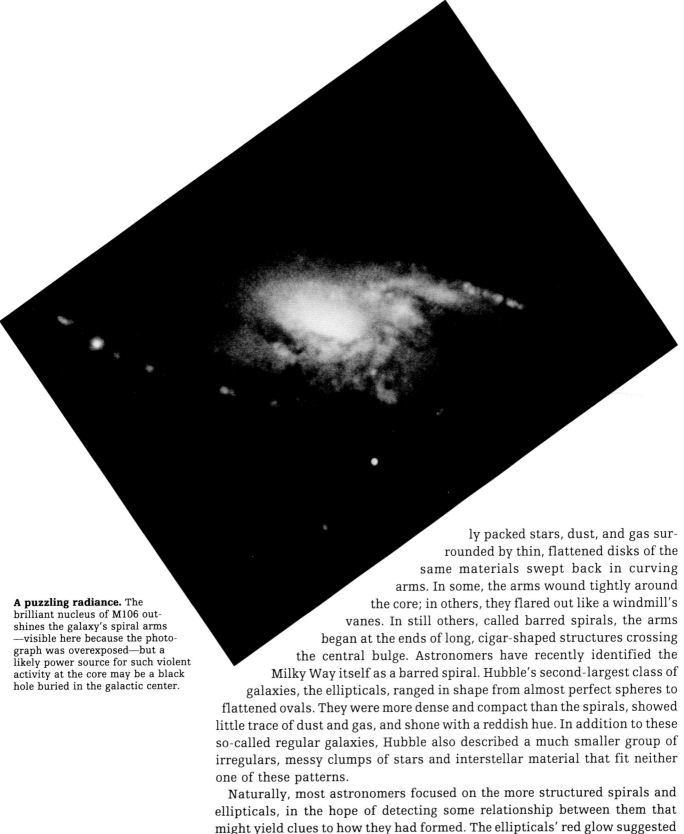

A puzzling radiance. The brilliant nucleus of M106 outshines the galaxy's spiral arms —visible here because the photograph was overexposed—but a likely power source for such violent activity at the core may be a black hole buried in the galactic center.

ly packed stars, dust, and gas surrounded by thin, flattened disks of the same materials swept back in curving arms. In some, the arms wound tightly around the core; in others, they flared out like a windmill's vanes. In still others, called barred spirals, the arms began at the ends of long, cigar-shaped structures crossing the central bulge. Astronomers have recently identified the Milky Way itself as a barred spiral. Hubble's second-largest class of galaxies, the ellipticals, ranged in shape from almost perfect spheres to flattened ovals. They were more dense and compact than the spirals, showed little trace of dust and gas, and shone with a reddish hue. In addition to these so-called regular galaxies, Hubble also described a much smaller group of irregulars, messy clumps of stars and interstellar material that fit neither one of these patterns.

Naturally, most astronomers focused on the more structured spirals and ellipticals, in the hope of detecting some relationship between them that might yield clues to how they had formed. The ellipticals' red glow suggested that most of their stars were older, and the absence of dust and gas seemed to indicate that, having exhausted their raw materials, they were no longer making new stars. Spirals, however, were dominated by bright young and middle-aged stars and had plenty of dust-and-gas fuel left to create new ones.

Some theorists speculated that ellipticals had formed earlier than spirals and represented a later stage in the spirals' evolution. Others argued that both types had formed at the same time but that ellipticals had started out with more massive stars (having coalesced, perhaps, out of denser pregalactic clouds) and so had evolved more rapidly. The latter theory did not preclude the possibility that spirals and ellipticals represent two phases of one basic type, but the prevailing view has become that inherent differences in initial conditions have led to two different evolutionary tracks. In any event, Hubble and others gathered enough evidence to convince most astronomers that, with the undeniable exception of the few irregulars, galaxies fell into two broad classes and a handful of standard variations.

ALTERNATE SCHEMES
The agreement was by no means universal. Some researchers had been focusing on other features that tended to cut across structural distinctions and might call for alternate typing approaches. In 1943, astronomer Carl Seyfert, who also worked at Mount Wilson, found about a dozen instances of a special type of galaxy that did not match any of the known models. Now known as Seyferts, these galaxies, usually spiral in shape, had extremely bright, almost starlike nuclei characterized by broad hydrogen emission lines in their spectra. (Spectral emission lines, like absorption lines, indicate the presence of certain elements, with broader lines denoting higher energies among the atoms, but they result from radiation emitted rather than absorbed by those elements.) These traits pointed to some kind of violent activity occurring at the core, though no one had any idea what it could be. Seyfert, who died tragically in an automobile accident a few years later, never learned the significance of his discovery, but astronomers would eventually recognize Seyferts as the first examples of an important category of star systems that were literally overflowing with energy: the active galaxies.

Shortly thereafter, Fritz Zwicky, a contentious and combative astronomer from the California Institute of Technology, further complicated the picture with a catalog of oddball galaxies, many of which seemed to be interacting with close companions. Zwicky's evidence suggested that, contrary to the general belief, galaxies did not live in splendid isolation but slipped and slid into each other, colliding, crashing, and sometimes changing the course of each other's evolution in the process.

The theory was certainly controversial, but no more so than the man himself. A Swiss national born in Bulgaria, Zwicky was a tall, imposing figure with piercing eyes, an acid tongue, and a complete disdain for all conventional ideas. Both his professional and his personal relations with fellow observers were so intense that several colleagues feared he might do them physical harm during heated arguments. Yet despite his abrasive personality, Zwicky was clearly a brilliant scientist and theoretician. Early on, he recognized that supernovae are a special phenomenon related to stellar evolution and realized that the best place to look for them was in other galaxies, where he found scores—adding to the previous total of twelve. He was also an early advocate of space exploration and, long before the Apollo program, described possible techniques for sending rockets to the Moon and bringing back samples of the soil for study.

His galaxy catalog, published in the mid-1950s, was the product of many years of observation and careful study of hundreds of photographs taken from the Palomar Observatory near San Diego—most notably with the world's new largest reflector, the 200-inch Hale Telescope. The catalog consisted of a detailed set of drawings that showed galaxies interacting in pairs and groups. The key features in the sketches were faint bridges, or links, that Zwicky had supposedly seen between many galactic companions and were just visible in some of the photographs. But Zwicky was hardly the most popular of figures, and few astronomers paid attention to his claims. He had grown so cantankerous and eccentric—at one time having ordered an assistant to fire a shot from a rifle along the line of sight of the Hale Telescope, apparently to clear the air and improve the view—that his telescope time had been restricted, and no one was eager to pursue his studies for him.

But the rich detail and strange morphologies being captured in Palomar images could not be ignored. By 1961, Soviet astronomer Boris Vorontsov-

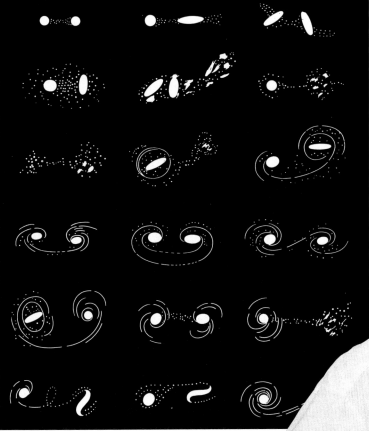

Astronomer Fritz Zwicky pioneered the study of interacting galaxies. In the 1950s, after extensive observations from Palomar Mountain, he prepared a set of sketches *(left)* to illustrate a variety of interactions, including faint bridges that seemed to connect some close neighbors. Zwicky theorized that the bridges were the debris of galactic encounters.

Velyaminov of the Sternberg Astronomical Institute in Moscow had found so many pairs of distorted and twisted galaxies in the National Geographic Society-Palomar Observatory Sky Survey—a complete photographic mapping of the entire Northern Hemisphere sky—that he was able to compile two catalogs of these strange objects, listing several hundred examples. Some of the variations included systems with corkscrew spiral arms, long star tails, and what appeared to be compact and brightly shining star-forming zones.

Fascinated by this evidence, Halton Arp, another Caltech researcher with a flair for the unconventional, gathered his own collection of weird galactic types from high-resolution Palomar photographs. Arp's 1966 photo album, appropriately entitled *Atlas of Peculiar Galaxies,* contained a bewildering array of warped structures that, like some cosmic Rorschach test, gave rise to a colorful list of names: the Antennae, the Playing Mice, the Telephone, the Carafe, the Fly's Wing.

Arp found more than 300 examples of interacting galaxies, so many that theorists could no longer dismiss them as mere curiosities. Obviously, Hubble's vision of a few grand classes of galaxies evenly spread throughout the heavens and occasionally passing in the night like celestial schooners was wrong. Close encounters were relatively common and were clearly an important force in the shaping of galaxies, perhaps accounting for many of the baffling structures of the irregulars. Particularly violent run-ins might even result in the annihilation of the weaker partner. Some 500 million years ago, for example, a near collision between the Milky Way and its two southern companions, the Large and Small Magellanic Clouds, dragged a narrow filament of hydrogen gas, known as the Magellanic Stream, from the two smaller galaxies toward the Milky Way's south pole. Eventually the Milky Way may swallow the clouds altogether.

Advances in radio and infrared astronomy, which gave scientists a much wider range of wavelengths to study, revealed more intriguing details and a possible relationship between interacting galaxies and the growing class of active galaxies such as the Seyferts. From early radio observations in the 1950s and 1960s, astronomers had noted the association of particularly strong radio signals with strangely shaped galaxies, which had perhaps been distorted by interactions in the past. In the 1970s, improved radio surveys indicated that enhanced emissions seemed most common from paired galaxies. By the 1980s, partly as a result of observations made by the Infrared Astronomical Satellite *(IRAS)* in 1983, strong support had developed for the view that encounters were the catalysts for many types of high-energy galactic events—including the formation of stars.

STAR BIRTHS BY THE HUNDREDS

Among their many attributes, infrared telescopes are especially effective at detecting young stars hidden within clouds of dust and gas: The clouds block visible radiation but are heated by the stars shining within and so radiate profusely at infrared wavelengths. Orbiting high above Earth's distorting atmosphere, *IRAS* captured an unparalleled amount of detailed information. It found hundreds of thousands of infrared sources during its ten months of observing, many of which coincided with the locations of galaxies and delineated compact regions of intense star formation. Some of these galaxies had been inconspicuous in optical surveys and were only now receiving their first serious attention. Others had already attracted notice because of regions of bright blue light within them, indicating a prodigious generation of stars. While relatively quiescent galaxies such as the Milky Way produce only a handful of stars each year, these so-called starburst galaxies were creating new stars at a hundred times that rate. Data from *IRAS* also confirmed that the shining hearts of many Seyferts were active star-forming areas. But most significant, a high percentage of all these systems had close companions.

The pieces seemed to fit. The breakneck production pace of starburst galaxies would quickly deplete all the raw material needed to make stars unless this birthing phase was only a fleeting episode, one short burst of intense activity in an otherwise long and placid life. And the most likely cause of such an episode, especially when there were other galaxies nearby, was a collision. Calculations have indicated that when two galaxies collide, or even if they pass close by, turbulent gravitational interactions can lead to an accelerated compression of interstellar gas and dust clouds, setting up perfect conditions for rapid star conception.

As plausible as this origin for starburst galaxies sounds, major questions remain. First, not every galactic collision generates starburst activity. Nor are all starburst galaxies part of interacting systems; they occasionally appear in total isolation. Thus, some theorists have argued that a starburst phase may be a normal part of every galaxy's evolution, perhaps akin to a teenager's sudden growth spurts. Yet this leaves unexplained the fact that starburst activity tends to be concentrated in relatively small areas of a galaxy—usually near its center—while normal star making is less localized and typically occurs in such places as a spiral galaxy's arms.

Despite their prevalence and their clear influence on some of the most

Sign of an encounter. Hydrogen linking two galaxies gives them their joint name, the Toadstool. The bridge, apparently the site of new star formation, seems to be a product of the galaxies' close interaction. Some astronomers believe that ne may have passed entirely through the ther, trailing cosmic rubble in its wake.

unusual formations and behaviors, galactic interactions cannot account for all the observed instances of active galaxies. Many are too alone in space to have been so dramatically influenced, and others found in groups exhibit none of the characteristic distortions that usually follow close encounters. What is to be made, for example, of the "cD" galaxies? These huge, bright objects have bulky elliptical bodies surrounded by far-flung envelopes of stars. (The term cD combines the standard astronomical designation c, for supergiant objects, with a D, for the diffuse nature of the envelope.) Among the largest galaxies ever observed, some measuring several hundred thousand light-years in diameter, they are found exclusively at the center of clusters of galaxies, outshining all their compatriots. A few show evidence of multiple nuclei, which may indicate that cD galaxies have swallowed other galaxies, thereby swelling to a gargantuan size.

Although they continue to have trouble explaining their discoveries, observers have managed to refine the classification of active galaxies, primarily on the basis of such features as the characteristic wavelengths of their radiation and the patterns of their spectral lines. Markarian galaxies, for example, first discovered by Benjamin Markarian and colleagues at the Byurakan Astrophysical Observatory in Armenia, radiate most intensely in the blue and ultraviolet portions of the electromagnetic spectrum. The category Seyferts now falls into two groups: Type 1 Seyferts possess the traditional broad emission lines, and those called Type 2 have narrower lines. Rather than representing two distinct kinds of galaxies, however, Type 1 and Type 2 Seyferts may reflect varying quantities of dust near the nucleus. According to this hypothesis, a Type 2 Seyfert would be one in which dust blocks radiation from rapidly moving gas clouds at the core, so that less turbulent gas farther from the center accounts for the narrower emission lines. A Type 1 Seyfert, on the other hand, would be one with little obscuring dust, so that radiation from its core dominates. Even more powerful than Seyferts are the N (for "nucleus") galaxies. Their tiny cores radiate so brilliantly that in relatively short photographic exposures, their outer envelope of stars completely disappears. Still another variation are the LINER (for "low-ionization nuclear-emission region") galaxies, exhibiting emission lines for elements that have been ionized, or stripped of electrons, to a minor degree. Unlike typical active galaxies, LINER galaxies radiate not from a

A celestial ring. This galaxy may have been a large spiral before one of the smaller galaxies at bottom passed through it. Astronomers believe the interloper's gravity drew stars and gas toward the center of the spiral. When the small galaxy moved on, the matter sprang out as a ring, festooned with new stars spawned by the shock.

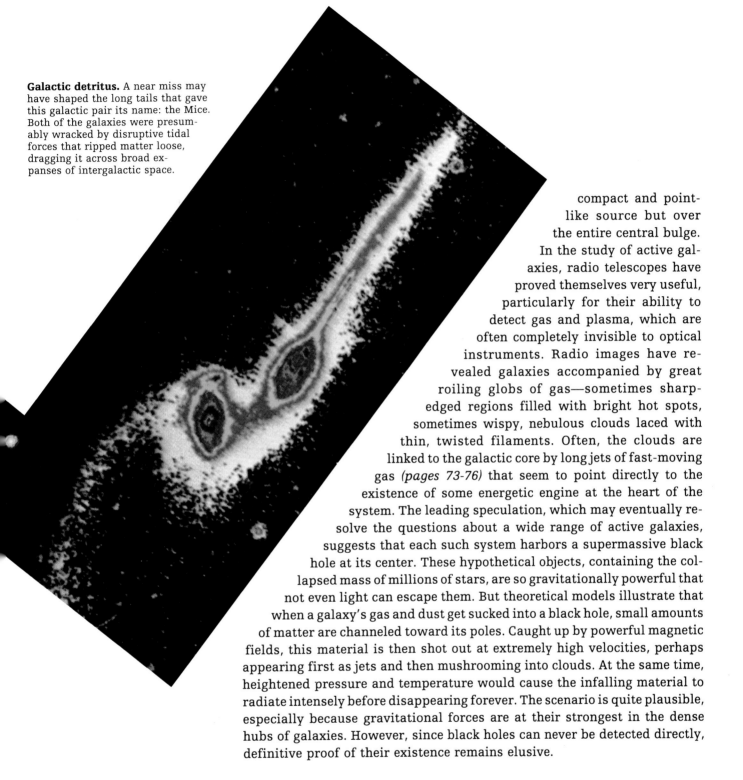

Galactic detritus. A near miss may have shaped the long tails that gave this galactic pair its name: the Mice. Both of the galaxies were presumably wracked by disruptive tidal forces that ripped matter loose, dragging it across broad expanses of intergalactic space.

compact and pointlike source but over the entire central bulge. In the study of active galaxies, radio telescopes have proved themselves very useful, particularly for their ability to detect gas and plasma, which are often completely invisible to optical instruments. Radio images have revealed galaxies accompanied by great roiling globs of gas—sometimes sharp-edged regions filled with bright hot spots, sometimes wispy, nebulous clouds laced with thin, twisted filaments. Often, the clouds are linked to the galactic core by long jets of fast-moving gas *(pages 73-76)* that seem to point directly to the existence of some energetic engine at the heart of the system. The leading speculation, which may eventually resolve the questions about a wide range of active galaxies, suggests that each such system harbors a supermassive black hole at its center. These hypothetical objects, containing the collapsed mass of millions of stars, are so gravitationally powerful that not even light can escape them. But theoretical models illustrate that when a galaxy's gas and dust get sucked into a black hole, small amounts of matter are channeled toward its poles. Caught up by powerful magnetic fields, this material is then shot out at extremely high velocities, perhaps appearing first as jets and then mushrooming into clouds. At the same time, heightened pressure and temperature would cause the infalling material to radiate intensely before disappearing forever. The scenario is quite plausible, especially because gravitational forces are at their strongest in the dense hubs of galaxies. However, since black holes can never be detected directly, definitive proof of their existence remains elusive.

THE QUASAR CONTROVERSIES

In the short but stunning history of radio astronomy, the prime attention getters have been the quasars—enigmatic objects whose discovery sparked fresh debates not only about active galaxies and black holes but also about the very dimensions of space.

In 1959, British radio astronomer Martin Ryle and others at Cambridge University produced the third edition of an atlas of all the radio-emitting

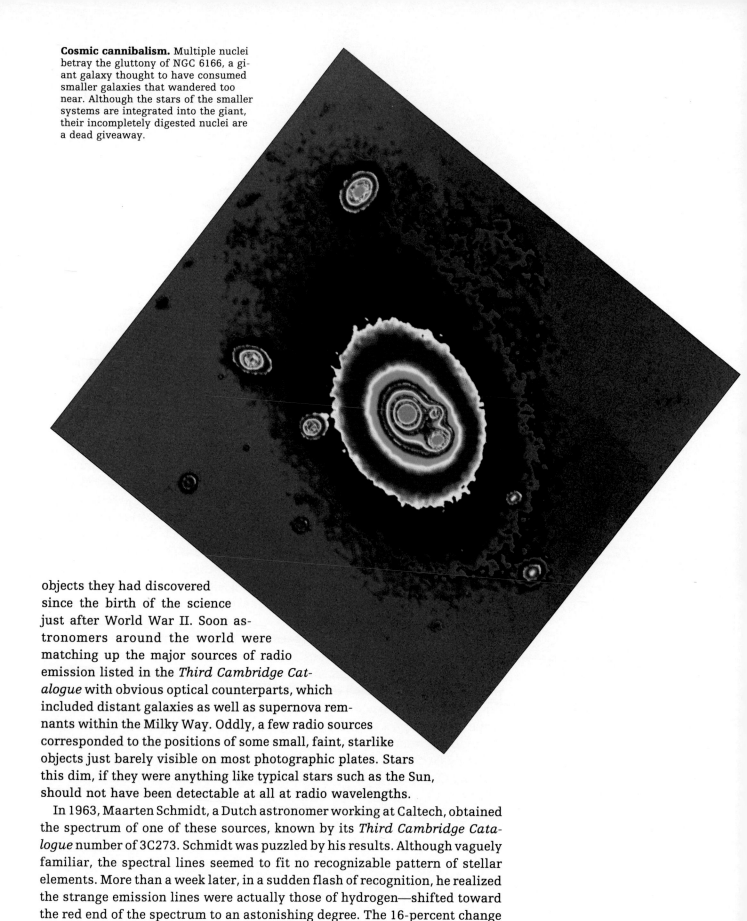

Cosmic cannibalism. Multiple nuclei betray the gluttony of NGC 6166, a giant galaxy thought to have consumed smaller galaxies that wandered too near. Although the stars of the smaller systems are integrated into the giant, their incompletely digested nuclei are a dead giveaway.

objects they had discovered since the birth of the science just after World War II. Soon astronomers around the world were matching up the major sources of radio emission listed in the *Third Cambridge Catalogue* with obvious optical counterparts, which included distant galaxies as well as supernova remnants within the Milky Way. Oddly, a few radio sources corresponded to the positions of some small, faint, starlike objects just barely visible on most photographic plates. Stars this dim, if they were anything like typical stars such as the Sun, should not have been detectable at all at radio wavelengths.

In 1963, Maarten Schmidt, a Dutch astronomer working at Caltech, obtained the spectrum of one of these sources, known by its *Third Cambridge Catalogue* number of 3C273. Schmidt was puzzled by his results. Although vaguely familiar, the spectral lines seemed to fit no recognizable pattern of stellar elements. More than a week later, in a sudden flash of recognition, he realized the strange emission lines were actually those of hydrogen—shifted toward the red end of the spectrum to an astonishing degree. The 16-percent change in wavelength translated on the redshift scale to a distance of three billion light-years, which meant the object was among the most distant ever detected and was thus far too bright to be a single star. In fact, for 3C273 to appear so bright at that range, it had to be at least 100 times brighter than a typical spiral galaxy, making it the most intrinsically luminous body in the known universe. Active galaxies were pale lights in comparison.

In the three decades since Maarten Schmidt's insight, scores more of these mysterious objects have been found and studied intensely, with an entire astronomical discipline springing up around them. Because the earliest ones were especially "radio loud," they were known as quasars (a shortened form of "quasi-stellar radio source"). Today, they are often referred to as QSOs, for "quasi-stellar objects." Most specialists describe quasars as the extremely bright and powerful nuclei of very distant (and thus very early) galaxies, perhaps driven at their centers by voracious, supermassive black holes.

THE REDSHIFT ISSUE
No matter what they might turn out to be, quasars attracted attention most of all because of their apparent extreme distance from Earth. If they are as far away as redshift measurements seem to indicate, then they are remnants of the universe's very earliest eras and would allow theorists, in effect, to travel back to those epochs. Not all astronomers see quasars as time machines, however. A small though vocal minority has argued that since some supposedly distant quasars seem physically associated with relatively nearby galaxies, the redshift rule may not apply universally to all types of extragalactic objects. Striking as it did at one of the central pillars of modern cosmology—the redshift evidence of an expanding universe—this hypothesis touched off what has been characterized as one of the most bitter episodes in the history of astronomy.

At the center of the debate is Halton Arp, the same astronomer who drew up an atlas of peculiar galaxies. Indeed, it was while investigating these extragalactic aberrations that Arp came upon what he believed was evidence for direct ties between some galaxies and quasars. Several Arp photographs show faint bridges apparently linking nearby galaxies with supposedly more distant quasars. Arp therefore argued that the high redshifts of these quasars were caused by factors other than distance; an extremely speculative theory proposed that the quasars had been ejected from their associated galaxy and that their high local velocity was distorting their redshift readings.

Most astronomers dismissed Arp's views out of hand, suggesting that the supposed connections were optical illusions produced by chance alignments. Some went so far as to impugn his integrity by remarking that most of the evidence of physical associations between objects of different redshifts came from photographs produced by Arp himself. A few eminent supporters, including the renowned astrophysicist Geoffrey Burbidge, made impassioned pleas for everyone to keep an open mind, but to no avail. In 1983, Arp was barred from the tools of his trade. Caltech's telescope allocation committee decided that his line of research was not worthy of support and that he would receive no more time for this work at the telescopes of the Mount Wilson and Palomar observatories. Arp refused to take up more conventional studies simply to please the committee; instead, he chose to leave Caltech for a position at the Max Planck Institute in Munich, where he continued to pursue his ideas. Referring to his abrupt and ignoble ouster, Burbidge later wrote, "No responsible scientist I know, including many astronomers who are strongly opposed to Arp's thesis, believes justice was served."

REACHING FOR THE EDGE OF THE UNIVERSE

Although Arp has been given little chance to prove them wrong, most modern astronomers assume that the quasars are not nearby objects that happen to be moving fast but, rather, extremely distant objects that may represent how the universe looked during its earliest epochs, perhaps only a billion years after the Big Bang.

While some quasars are associated with clusters of galaxies—at comparable redshifts, unlike Arp's peculiar examples—most appear to be much deeper in space than any observed galaxies. Maarten Schmidt, who first identified these objects, also made a remarkable discovery about their nature: They seemed to be most abundant at redshifts of about 2, equivalent to a distance of approximately 12 billion light-years. (The value of 2 derives from a standard formula in which the laboratory-measured rest wavelength for a key emission line is subtracted from the redshifted wavelength and the total is divided by the rest wavelength.) The fact that there are none closer than two billion light-years and progressively more at points farther out suggested that even more quasars should be seen at even greater distances. But by the early 1980s, only one or two had been seen beyond a redshift of 3. There seemed to be a redshift limit beyond which quasars could not be seen.

"The limit implied that for at least one type of object astronomers were seeing to the edge of the universe," wrote Kitt Peak astronomer Patrick Osmer in 1982. "It looked as if quasars had formed suddenly in a great burst of activity." If quasars are really the young, bright hearts of faint, nascent galaxies, then the redshift cutoff might correspond to an era when all the galaxies were created. The bunching of quasars around a single point in time was strong support for the hypothesis that all galaxies formed very quickly and nearly simultaneously in a sudden, brilliant bloom.

Less than five years later, however, Pat Osmer would be among a host of observers vying for the distance record in the discovery of quasars. Between August 1986 and August 1991, more than ten quasars were found with redshifts greater than 4. Appropriately, Maarten Schmidt is a member of the team that has spotted half of them, including the most distant object ever seen—a quasar at redshift 4.897. Discovered in early 1991, quasar PC 1247 + 3406 is receding at a rate equivalent to a distance of 14 billion light-years.

"If the age of the universe is 15 billion years, this quasar was emitting light just under 1 billion years after the Big Bang," says Donald Schneider of

Princeton, a codiscoverer with Schmidt and fellow Princeton astronomer James Gunn, "which places some serious constraints on theories of galaxy formation." Suddenly, the theorists seemed to be running out of time for the standard schedule of events that were supposed to have led to the creation of galaxies. If those most distant quasars had somehow formed at an accelerated rate, they showed no sign of such behavior. As Schmidt put it, "One of the interesting things about our redshift 4.897 quasar is how normal it is. Except for its great distance, it shows no significant differences from other quasars." Debate continues as to the implications of these long-distance beacons, whether they call for revisions in the generally accepted time scale of the universe, demand new explanations for the origins of galaxies—or, perhaps, lend support to Arp's challenge. Quasars may, in a sense, be receding back into the depths of mystery.

EYES ON THE INVISIBLE

While many researchers focus on the quasar's role in pushing back the edges of the known universe, others look at quasars from an entirely different perspective. "Even if one has no interest in quasars or little faith that we will ever understand their genesis," Pennsylvania State University astronomer

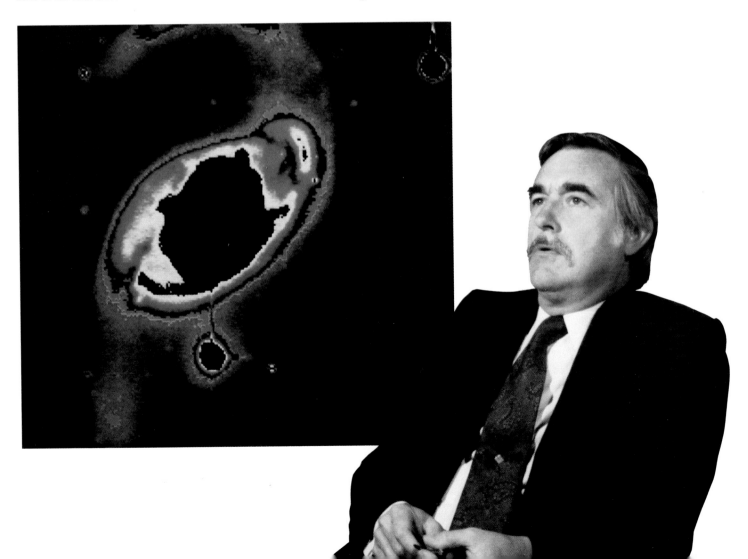

A heretic's evidence. In a photograph by controversial astronomer Halton Arp *(below, right)*, a large spiral galaxy located relatively near the Milky Way and a quasar widely assumed to be a billion light-years more distant appear to be physically linked by a bridge of matter. Arp contends that pictures such as this support his theory for the noncosmological nature of quasar redshifts: He believes that the high redshifts may be caused by something other than increasing distance resulting from the expansion of the universe.

and quasar studier Daniel Weedman has written, "they remain fundamentally useful as probes of everything between us and them." Because the light from quasars has been traveling across space for billions of years, he notes, it has been affected and changed by all the dust, gas, and interstellar molecules it has encountered along the way to Earth. As a result, quasars may prove to be an important tool in the continuing search for the hidden components of the universe.

According to Weedman, who has extensively investigated alternate uses for quasar data, "Quasars are background searchlights which silhouette otherwise dark or invisible material." They serve as cosmological probes in two ways: first, through special instances of the unusual optical phenomenon known as gravitational lensing, and second, through absorption-line patterns in their spectra caused by intervening material.

The principle behind gravitational lensing is that massive foreground objects such as galaxies generate strong gravitational fields that can actually bend, or refract, the light from distant background objects such as quasars. If a gravitational lens was at work, an observer on Earth would see not only the real quasar but also an illusionary double or, perhaps, multiple false images, including ring- or arclike structures. Although predicted by Albert Einstein early this century, the first example of a gravitational lens—which turned out to be a visible galaxy that was distorting the image of a quasar along the same general line of sight—was not found until 1979. Since then, several well-publicized "multiple quasars" have been identified with no obvious candidate lenses to account for them, leaving open the possibility that the responsible agents could be enormous, invisible clouds of dust and gas. As observational techniques have improved, astronomers have discovered that careful measurements of slight time delays in the arrival of light from the different images—that is, from the real object and its refracted version—can provide fairly accurate estimates of the size and mass of the intervening body. Some examples support the existence of very large accumulations of matter, equivalent in mass to hundreds of normal galaxies.

The second technique, based on absorption-line patterns, is more time-consuming and tedious, for quasar spectra must be obtained with long exposures, then carefully analyzed to separate out those dark lines caused when the quasar's light passes through a cooler intervening gas cloud. Diligent study has produced one most intriguing result: the discovery of small clouds of almost pure hydrogen. Hitherto unseen, these miniature clouds are widely dispersed and, more surprising, seem never to have been associated with any galaxy or cluster of galaxies, which have occasionally been observed to spin off such wispy patches. Could they be leftovers from the birth of the universe? Or are they transient phenomena, coming and going throughout cosmic history, like clouds on a summer day?

According to some theorists, these clouds of intergalactic material may represent a class of objects that tried—and failed—to develop into coherent, observable structures. Gregory Bothun, an astronomer at the University of

Dutch-born astronomer Maarten Schmidt, shown here with a device for making fine measurements on astronomical photographs, was the first to decipher, in 1963, the bizarre spectrum of 3C273, a starlike object that was given the name quasar—for "quasi-stellar radio source."

Oregon, thinks the universe could be teeming with large, dark, dull objects that progressed no farther than the first halting steps toward galaxy formation approximately 10 billion years ago. Distinct from the protogalaxies pursued by observers such as Haynes and Giovanelli, these failed galaxies did not challenge the notion of one distinct era of galaxy creation. But, in analogous fashion to brown dwarfs, they did suggest that the process could sometimes fall short.

The first example of what Bothun calls low-surface-brightness galaxies was discovered accidentally in 1987, when he, Chris Impey of the University of Arizona, and astrophotography wizard David Malin of the Anglo-Australian Observatory were looking for faint dwarf ellipticals. A special light-enhancing photographic technique developed by Malin picked up an unusually large object with a slightly fuzzy nucleus and a faint outer envelope in the vicinity of the constellation Virgo. Later observations revealed Malin 1 to be almost eight times bigger than the Milky Way. Lying some 715 million light-years from Earth, Malin 1 is nearly invisible because most of its mass exists in the form of diffuse hydrogen gas. In fact, the system's minuscule component of stars is so scattered that the galaxy as a whole is little brighter than the surrounding sky.

In early 1989, Bothun and his coworkers discovered a second, smaller example. In both cases, the large spherical bulge consisted of very old stars, as was typical for most galaxies, but neither object had continued the growth process to form a surrounding disk. According to Bothun, these objects had apparently used up a critical portion of their original density in forming their first stars and then had essentially shut down. He believes such quiescent, nonevolved galaxies could number somewhere in the billions. His argument has an appealing logic. Astrophysicists know that stars form from clouds of gas with an efficiency of about 10 percent, leaving 90 percent of the cloud's matter uncondensed. Since galaxies themselves are thought to have coalesced in a similar manner, there seems to be, in effect, a theoretical mandate for the existence of Bothun's dim failures.

THE DARK SECRET OF THE UNIVERSE
The idea that there may be more to the universe than meets the eye was not new. As early as the 1930s, Dutch astronomer Jan Oort carefully measured the positions and motions of certain stars outside the disk of the Milky Way and calculated how much mass would be needed to produce their observed motions according to the laws of physics. Unfortunately, when he estimated the mass of all the visible stars in the galaxy, his calculations came out about 50 percent short. Only by adding a correction factor—what he assumed must be faint, uncounted stars—could he generate enough mass to explain the observed gravitational effects.

At about the same time, Fritz Zwicky was noticing that the velocities of individual galaxies within certain groups were so great that they should not have been gravitationally bound to the other members of the group. (Redshift data, in addition to revealing the overall recession of galaxies, can also be used to measure a galaxy's local motions—its spin, for example, or its orbital relationship with nearby galaxies.) Zwicky calculated that the amount of mass needed to counterbalance the centrifugal force created by these high velocities was nearly 100 times larger than he could see at optical wavelengths. The prescient Zwicky proposed that some dark, unseen material must be serving as intergalactic glue.

Further evidence backing Zwicky's suggestion came in the 1970s from Vera Rubin, of the Carnegie Institution of Washington, D.C., who had been studying the curious rotation rates of spiral galaxies. Astronomers had long inferred that the distribution of mass in a galaxy was indicated by the distribution of its luminosity; for example, a spiral galaxy's bright central bulge would contain most of the system's mass, with the rest thinly scattered in the

A Collection of Cosmic Jets

Among the strangest and most powerful phenomena in the universe are cosmic jets—streams of electrically charged, superheated gas that bore through space at extraordinary speeds, marking bright paths often more than a million light-years in length. Current theory holds that these cosmic blowtorches originate in the whirling cores of protostars or in the black holes—millions to billions of Suns in mass—that may lurk at the centers of certain especially energetic galaxies, including quasars.

Whether the jets are stellar or galactic, the process of formation is the same. It begins as gravity funnels gases into a rotating disk around the protostellar core or black hole. From the disk's inner rim, convective energy and radiation pressure beam a percentage of the spiraling matter back into space. Initially accelerated to tremendous speeds, the high-pressure stream is eventually slowed by the gas and dust of space, culminating in a so-called shock, behind which the jetted material collects like dammed-up water. In the case of the more powerful galactic jets, the accumulated gas forms enormous, optically invisible lobes that emit prodigious amounts of radio noise.

Detection of such radio lobes is one way scientists infer the existence of these cosmic streams. Although jets are believed to occur in pairs, emanating from opposite ends of the protostar's or galaxy's rotational axis, often only a single jet is detectable by present means. The unseen partner may be rendered invisible because it is racing away from Earth at speeds approaching that of light. Jets frequently assume odd shapes—needle thin, right-angled, or plumed—as dictated, perhaps, by differences in the jet "engines" or by capricious intergalactic winds.

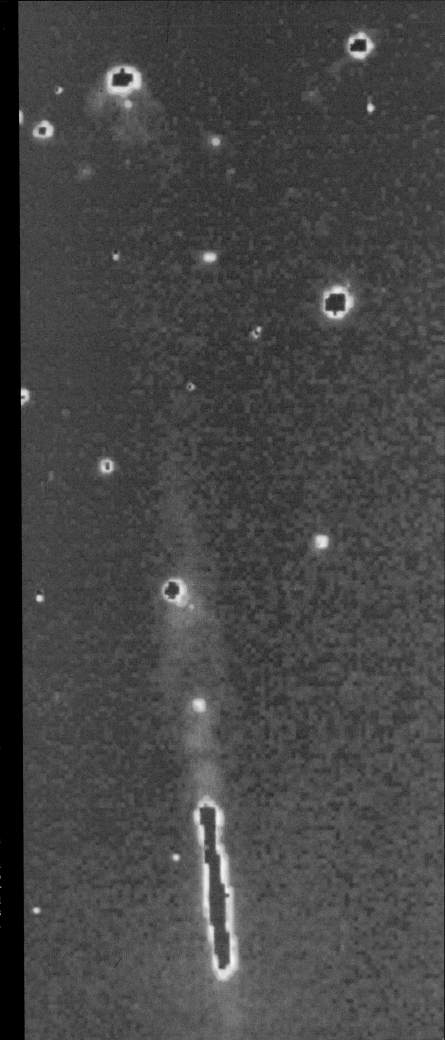

Blazing like an acetylene torch across six trillion miles, a jet of gas *(pink, light blue)* shoots out of the heart of unseen infant star HH111 *(bottom center of image)* in the constellation Orion. Along the jet's cooler blue extension, two hot spots, called Herbig-Haro objects *(center and top)* after the astronomers who discovered them, may mark shocks where the protostar's beam twice collided with dense interstellar gas.

Giant radio-emitting lobes of gas that flank galaxy Centaurus A *(far left, red contours)* dwarf the 15,000-light-year-long jet emanating northward from the galactic center *(small red circles)*. No jet has yet been detected feeding the southern lobe. Centaurus A's peculiar visible structure—an elliptical galaxy with an embedded dust lane—suggests that the galaxy may have consumed a smaller star system and is disgorging jets of gas as a result.

As luminous as 40 million Suns at its brightest point, a lone jet *(green)* 5,000 light-years long bursts flamelike out of elliptical galaxy M87's nucleus (blackened here to enhance the jet's visibility), as shown in this image—a composite made at near-ultraviolet and optical wavelengths—of the center of the Virgo cluster. Observations have revealed that the entire jet has varied a great deal in brightness over the last thirty years—a so-far-unexplained mystery.

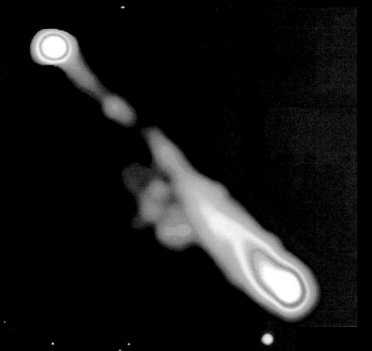

Across three billion light-years, quasar 3C273's jet radiates as much light as the Milky Way galaxy and a million times as much radio noise. (Light regions are the strongest emitters; darker regions, the weakest.) The jet's great velocity—185,800 miles per second—and its very peculiar head-on orientation relative to Earth make it appear to travel eight times faster than light.

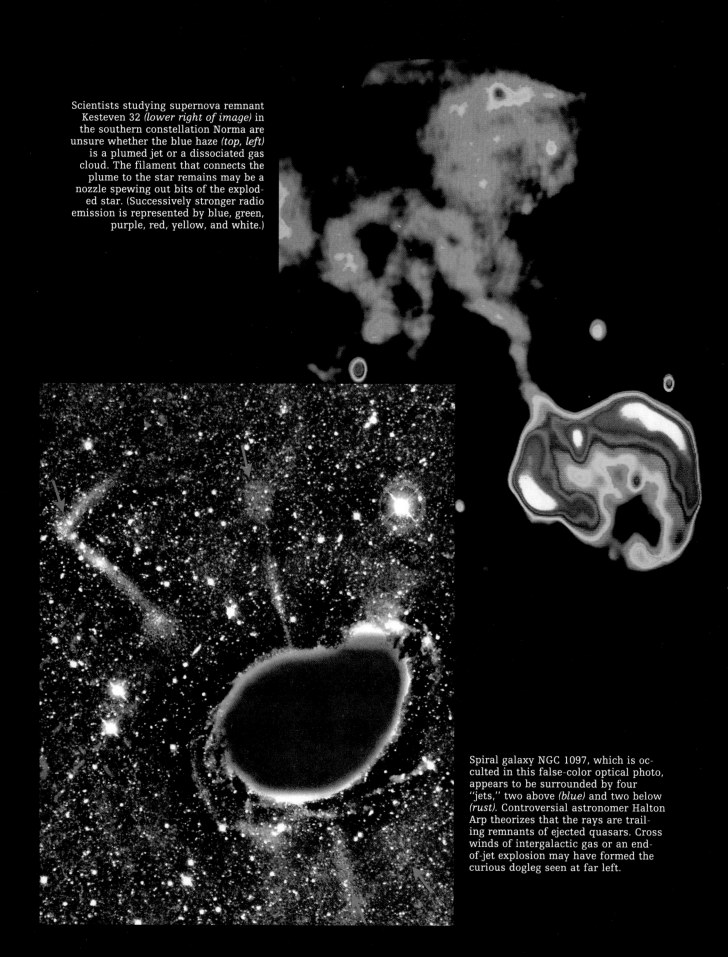

Scientists studying supernova remnant Kesteven 32 *(lower right of image)* in the southern constellation Norma are unsure whether the blue haze *(top, left)* is a plumed jet or a dissociated gas cloud. The filament that connects the plume to the star remains may be a nozzle spewing out bits of the exploded star. (Successively stronger radio emission is represented by blue, green, purple, red, yellow, and white.)

Spiral galaxy NGC 1097, which is occulted in this false-color optical photo, appears to be surrounded by four "jets," two above *(blue)* and two below *(rust)*. Controversial astronomer Halton Arp theorizes that the rays are trailing remnants of ejected quasars. Cross winds of intergalactic gas or an end-of-jet explosion may have formed the curious dogleg seen at far left.

much dimmer arms. Physicists also assumed that a galaxy's outermost stars would be orbiting much more slowly than those nearer the hub, just as the inner planets revolve around the Sun much faster than do the sluggish outer planets. But contrary to expectation, Rubin was finding, in virtually every galaxy she studied, that high orbital velocities continued to the galaxy's edge. This led her to conclude that outer stars were being sped up by the gravitational attraction of unseen matter lurking in an extended halo well beyond the galaxy's visible dimensions. "Nature has played a trick on astronomers," says Rubin. "We thought we were studying the universe; now we know we are studying only the small fraction that is luminous."

Radio observations of shells of hydrogen at extended distances from the cores of galaxies have given some support to the idea of dark-matter halos. Furthermore, space-based x-ray observatories have revealed previously invisible massive clouds of hot gas around elliptical galaxies. Even larger, more diffuse patches of x-ray-emitting gas have been found in the supposedly empty spaces among groups of galaxies. And the discovery of Malin 1 has opened up the possibility of extensive congregations of dark matter unrelated to any visible galaxies.

Despite its elusiveness, dark matter continues to fascinate scientists, in large part because it could determine the fate of the universe. The Big Bang theory leaves unanswered at least one major question: Will the observed expansion continue forever, creating a so-called open universe, or is there so much matter in the cosmos that its total gravitational force will eventually arrest the outward motion and start pulling everything back in to a final Big Crunch? An abundance of dark matter—at least ten times as much as the luminous variety—would tip the scales to the latter, closed universe.

THE CLUSTER EFFECT
Dark matter may also play a critical role in a related puzzle concerning the overall structure of the universe. Almost from the moment of its conception, Hubble's vision of galaxies rushing outward from a single point in time and space produced a troubling paradox. If the universe was steadily and evenly expanding in all directions, why did its visible elements look so clumpy? Although theorists had carefully worked out how galaxies themselves could come into being under the influence of gravity, they were unable to explain the massive associations that were becoming more apparent with each advance in observational techniques.

Once again, the controversial figure of Fritz Zwicky was at center stage. His work on interacting galaxies in the 1930s and 1940s had persuaded him that galaxies tended to congregate not only in pairs and small groupings but also, at times, in even larger clusters. He was among the first to discover that the Milky Way is part of a small bunch of some twenty galaxies that he called the Local Group, distributed over an area about four million light-years wide. Gravitational attraction among close galactic neighbors was strong enough, he argued, to keep them together as a group even as the universe

stretched apart. The perceptive Zwicky also believed larger, possibly super-size clusters, each with hundreds and even thousands of member galaxies, existed deeper in space.

Zwicky's reputation was such that his ideas were bound to be challenged purely on principle, even if the evidence in his favor was strong. Astronomers quickly split into two distinct camps over the question of large-scale structures. Understandably, Edwin Hubble staunchly defended his concept of uniform expansion, with galaxy distribution smooth and free of embarrassing "lumps." But in the 1950s, a feisty young French astronomer named Gérard de Vaucouleurs discovered, while revising a standard galaxy catalog, that the Milky Way and its Local Group neighbors seemed perched at the edge of an even larger grouping of galaxies. Fresh from Paris and the Sorbonne—and somewhat removed from the American debate over celestial lumpiness—Vaucouleurs proposed that the Local Group was actually just one part of a gigantic gathering of galaxies that included another comparable cluster in the general direction of the constellation Virgo, as well as some thirty other smaller systems.

Then, in 1958, George Abell of Caltech completed a painstaking examination of the 3,000 glass plates of the monumental National Geographic-Palomar Sky Survey, identifying more than 2,700 clusters, which he defined as a minimum of fifty bright galaxies inside a circle with a radius of 6.5 million light-years. Yet a case could not be based on apparent associations in two-dimensional photographic surveys if it was to be thoroughly convincing. The third dimension, distance, was crucial to any linkage, and distance depended on redshift data.

At the time, measuring redshifts was an even more laborious task than peering at tiny smudges on photographic plates. Hubble's assistant Milton Humason, a dedicated observer who began his career driving mules for the

Inexplicable structures. The strand of bright galaxies strung across this photograph of the sky belongs to the Perseus cluster, part of a much larger agglomeration known as the Perseus-Pisces-Pegasus supercluster. In the diagram at right, which plots galaxies according to their redshifts as measured from Earth, the distribution of the portion of the Perseus cluster visible above is marked by a narrow red band; the rest of the supercluster stretches to the right. By some estimates nearly a billion light-years across at its fullest extent, the supercluster contravenes prevailing theories for the evolution of the universe, which do not allow for the formation of such massive structures.

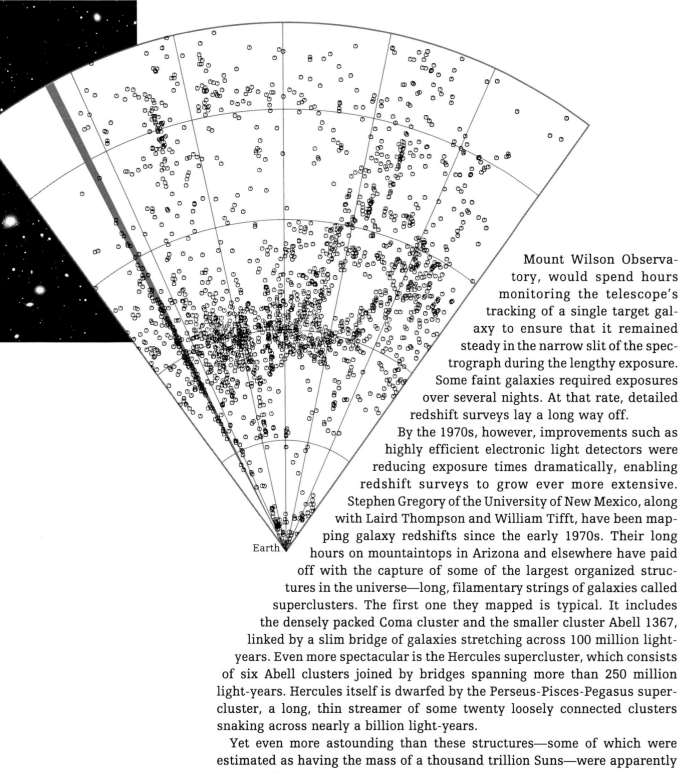

Mount Wilson Observatory, would spend hours monitoring the telescope's tracking of a single target galaxy to ensure that it remained steady in the narrow slit of the spectrograph during the lengthy exposure. Some faint galaxies required exposures over several nights. At that rate, detailed redshift surveys lay a long way off.

By the 1970s, however, improvements such as highly efficient electronic light detectors were reducing exposure times dramatically, enabling redshift surveys to grow ever more extensive. Stephen Gregory of the University of New Mexico, along with Laird Thompson and William Tifft, have been mapping galaxy redshifts since the early 1970s. Their long hours on mountaintops in Arizona and elsewhere have paid off with the capture of some of the largest organized structures in the universe—long, filamentary strings of galaxies called superclusters. The first one they mapped is typical. It includes the densely packed Coma cluster and the smaller cluster Abell 1367, linked by a slim bridge of galaxies stretching across 100 million light-years. Even more spectacular is the Hercules supercluster, which consists of six Abell clusters joined by bridges spanning more than 250 million light-years. Hercules itself is dwarfed by the Perseus-Pisces-Pegasus supercluster, a long, thin streamer of some twenty loosely connected clusters snaking across nearly a billion light-years.

Yet even more astounding than these structures—some of which were estimated as having the mass of a thousand trillion Suns—were apparently gaping holes between them. As Gregory and his colleagues were totting up the galaxies in the Coma cluster, for example, they noticed a vast empty region between it and the Milky Way's Local Group. At first, the gap seemed only a strange quirk, peculiar to one region of the sky. Then they found another vacancy in the foreground of the Hercules cluster. In fact, almost every astronomical search for superclusters has also uncovered large areas of mysteriously empty space.

In 1981, during an attempt to determine the average density of the universe by counting galaxies in selected areas of the sky, Robert Kirshner and co-workers at the University of Michigan discovered an eerily empty region 250 million light-years wide near the constellation Boötes. "We thought Boötes was a nice, boring section that represented a typical piece of the universe," says Kirshner, who later moved to Harvard. "Discerning the large-scale structure of the universe was not our intention." But the Boötes void, as Kirshner called it, would soon take its place as the most obvious example of the universe's surprising patchwork quality.

BUBBLES AND VOIDS
Throughout the 1980s, the Harvard-Smithsonian Center for Astrophysics has led efforts to map the three-dimensional distribution of galaxies. In 1985, astronomers Margaret Geller and John Huchra began a significant extension of earlier Harvard surveys by looking at the redshifts of galaxies approximately two times deeper in space. Their approach involved delineating a pie-shaped wedge of sky that was 117 degrees across and 6 degrees from top to bottom—its point anchored at the Whipple Observatory in southern Arizona—then precisely determining the position of each galaxy falling within the slice.

While plotting the combined celestial coordinates and redshift distances as points on a three-dimensional computer map, graduate student Valerie de Lapparent noticed that the galaxies seemed to form huge structures resembling sheets wrapped around great blank areas. Careful analysis by Geller, Huchra, and Lapparent confirmed that the galaxies were indeed arrayed as if over the surfaces of gigantic bubblelike voids, some as large as 100 million light-years in diameter. What had once seemed to be long stringy clusters of galaxies were now being interpreted as parts of bubble surfaces, their flattened, or filamentary, appearance caused by the angle of view from Earth. And the Boötes void, which was not covered in the survey, appeared to be a common phenomenon.

The subsequent announcement of a bubble universe shook the astronomical world. The observed features posed serious challenges to almost every existing model for the generation of large-scale cosmic structure. For example, the gravitational forces invoked by Zwicky to explain small-scale clustering could not have produced such sharp distinctions between galaxy-rich bubble surfaces and immediately adjacent voids. Rather, the voids looked like spaces blown clean by mammoth explosions.

Actually, a few years earlier, Princeton's Jeremiah Ostriker and Lennox Cowie of the Johns Hopkins University had proposed a highly speculative theory that extremely massive stars, much larger than any now known, had formed during the earliest phases of the universe. These megastars had rapidly advanced to the supernova stage, exploding by the millions and

The Great Wall. Located in the opposite part of the sky from Perseus-Pisces-Pegasus, the so-called Great Wall is among the largest coherent structures yet discovered in the universe. Shown here from two computer-generated perspectives, the Great Wall defies theoretical efforts to explain its great size. Astronomers believe that the voids it outlines, each up to 150 million light-years across, may be as important to understanding the structure of the universe as they are to understanding the structure of the Great Wall itself.

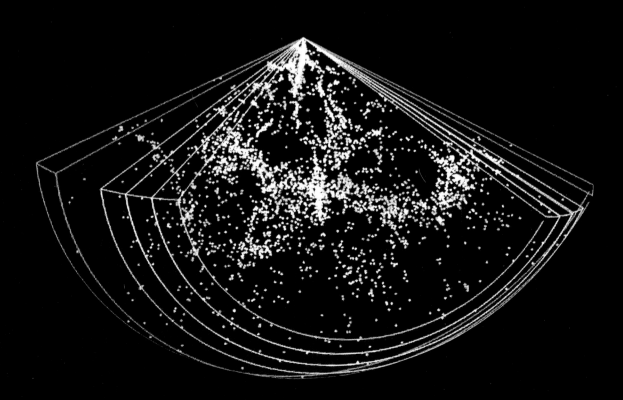

pushing away the remaining clouds of gas, which would eventually evolve into galaxies. This scenario seemed to jibe with the Harvard observations, but no combination of forces was great enough to create bubbles of the size being detected, no matter how many explosive charges—including the eruption of quasars and entire primordial galaxies, as well as supernovae—were added to the theoretical model. Other observers felt that, while the voids were really there, the bubble description was misleading. Stephen Gregory, among others, has asserted that the universe may be more like a sponge, in which both the solid portions and the air passages, or the voids, are connected. The shaping mechanisms would undoubtedly be different, but so far they have proved only marginally less problematic than those for a soap-bubble universe of separated voids.

Inevitably, attention has focused on the possible influence of dark matter. According to Robert Kirshner, "We know for sure that the Boötes void is not filled with big, bright galaxies like ours. We're now investigating the possibility there might be a somewhat different population of galaxies in the void." If it turns out that empty regions are actually teeming with undetected galaxies of dark matter, preserving the notion of a relatively even spread of materials throughout the universe, cosmologists would still have to explain why dark matter and light matter separated into such distinct structures. Some theorists have put the equivalent of a question mark into their equations, suggesting that unknown exotic particles make up the dark matter and act in ways that, almost too conveniently, account for the current state of the universe. Others are working on the notion that the mere presence of that much additional matter would have made gravity a stronger force throughout the universe's development, capable of turning minor density fluctuations shortly after the Big Bang into the present observed pattern of dense webs of galaxies around less dense regions dominated by gas and dust. But much more will have to be learned about what dark matter is and how it actually arrays itself across the cosmos before more definitive conclusions can be drawn.

BEHOLD, THE WALL

In the meantime, Geller and Huchra have continued to cut additional slices from adjacent regions of the sky, aiming to obtain redshifts for some 15,000 galaxies. The prominent bubble structures have carried over from one section to the next. And stretching from edge to edge across the four wedges plotted by the end of 1989 was a truly monstrous sheet of galaxies the astronomers dubbed the Great Wall.

Its dimensions were awesome. A half-billion light-years long, some 200 million light-years wide, and a relatively skinny 15 million light-years thick, it was by far the most imposing cosmic structure ever observed. Nevertheless, Geller and Huchra knew that its preeminence was not likely to endure for very long. "The size of the largest structures we detect is limited only by the extent of our survey," they wrote in announcing the discovery. Theorists must therefore wrestle with an apparently even clumpier universe—and face the haunting possibility that next week's news will leave their latest speculations out of date.

In another part of the sky, other searchers have finally cornered a less clearly structured but no less intriguing cosmic behemoth that also has the theorists scrambling. Early in the 1980s, a team of seven scientists, whose chief spokesperson is Alan Dressler of the Carnegie Institution's Mount Wilson and Las Campanas observatories, undertook an extensive mapping of galaxy motions in the vicinity of the Milky Way. The initial findings of the group, known to supporters and detractors alike as the Seven Samurai, confirmed earlier studies that had detected local motions diverging from the general recession, or Hubble flow, of galaxies. These so-called peculiar motions were thought to be caused by gravitational tugging from nearby clusters. The Seven Samurai's data showed that the Local Group of galaxies was streaming toward the Virgo cluster, and that the Local Supercluster—which includes the Local Group and the Virgo cluster—was not keeping up with the Hubble flow, pulled sideways toward the next nearest supercluster, Hydra-Centaurus.

So far, the result was not very startling: If Hydra-Centaurus was more massive than the Local Supercluster, then its greater gravitational force would naturally be pulling the Local Supercluster toward it. But Dressler and his fellow scientists soon found that Hydra-Centaurus was itself also moving. Together, the two superclusters—encompassing many thousands of galaxies—were rushing toward what had to be an even greater mass at velocities up to 700 miles per second, the largest deviation from the Hubble flow ever recorded. Survey maps seemed to indicate a profusion of galaxies more or less in the direction of these peculiar motions, a super-supercluster estimated to lie approximately 150 million light-years from the Milky Way. "I called it the Great Attractor," says Dressler, "because a structure so gargantuan deserved an equally grand name."

Although many astronomers were initially skeptical, Dressler announced clinching evidence in January 1990. The team's surveys had reached a point where galaxies no longer showed peculiar motion, and farther out were galaxies with an opposite flow, back in the direction of the Milky Way. The Seven Samurai had reached the heart of the Great Attractor. While the discoverers could relish the success of a masterful piece of detective work, cosmologists would have to deal with yet more evidence of the universe's ability to create supposedly impossible concentrations of matter.

But so it has been throughout the brief history of galaxy studies. Theorists have found themselves continually driven by the unfolding diversity of the island universes and their associations. The best minds can only wonder what new mysteries and challenges await as advanced observational instruments probe with an ever-keener vision. Given the perplexing and often surprising course of the story so far, the geniuses of the next generation of galactic astronomers may be those most open to the unexpected.

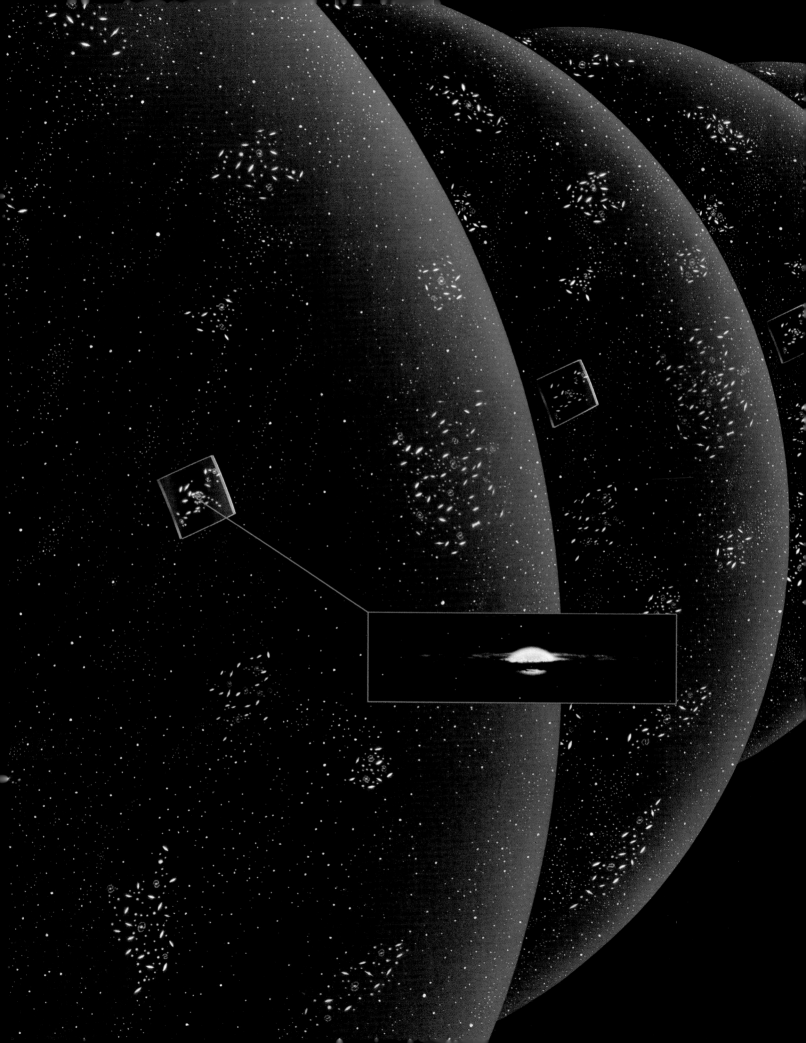

MYSTERIOUS MOTIONS

With the publication in 1929 of his findings that galaxies everywhere are hurtling away from the Milky Way at speeds directly proportional to their distance, astronomer Edwin Hubble touched off a cosmological revolution. The largely static cosmos of Kepler and Newton was replaced with a universe in which the very stuff of space is pushing apart everything it holds. The effects of the cosmic expansion can most easily be visualized by imagining the universe as the surface of a balloon sprinkled with galactic glitter. (Areas inside and outside do not exist in this visualization.) The surface has no center or edge, as is the case with the actual universe. From the vantage of the Milky Way *(inset, left)*—or from any other vantage—all galaxies on the balloon universe appear to recede as the balloon itself grows in size.

The most radical implication of the expansion, or Hubble flow, is that the universe was once a single point, out of which all space, time, energy, and matter exploded. In the eons following the Big Bang, outflying matter condensed under gravity's pull, first into stars, then galaxies, clusters of galaxies, and clusters of galaxy clusters, or superclusters—and finally, vast galactic filaments stretching a billion light-years or more. During the 1970s, astronomers found that individual galaxies exert subtle gravitational influences on each other, causing "peculiar motions," localized movements that are at times counter to the cosmic expansion. But beyond the galactic level, concentrations of mass were believed to be more or less equally distributed. Attractions were thus thought to balance each other out, leaving pure Hubble flow. Today this view is being challenged. Observations reveal that the Milky Way and all the galaxies within a 200-million-light-year region of space are rushing toward a distant attractor at better than a million miles per hour. Understanding this colossal peculiar motion may prove the key to unlocking the fate of the cosmos.

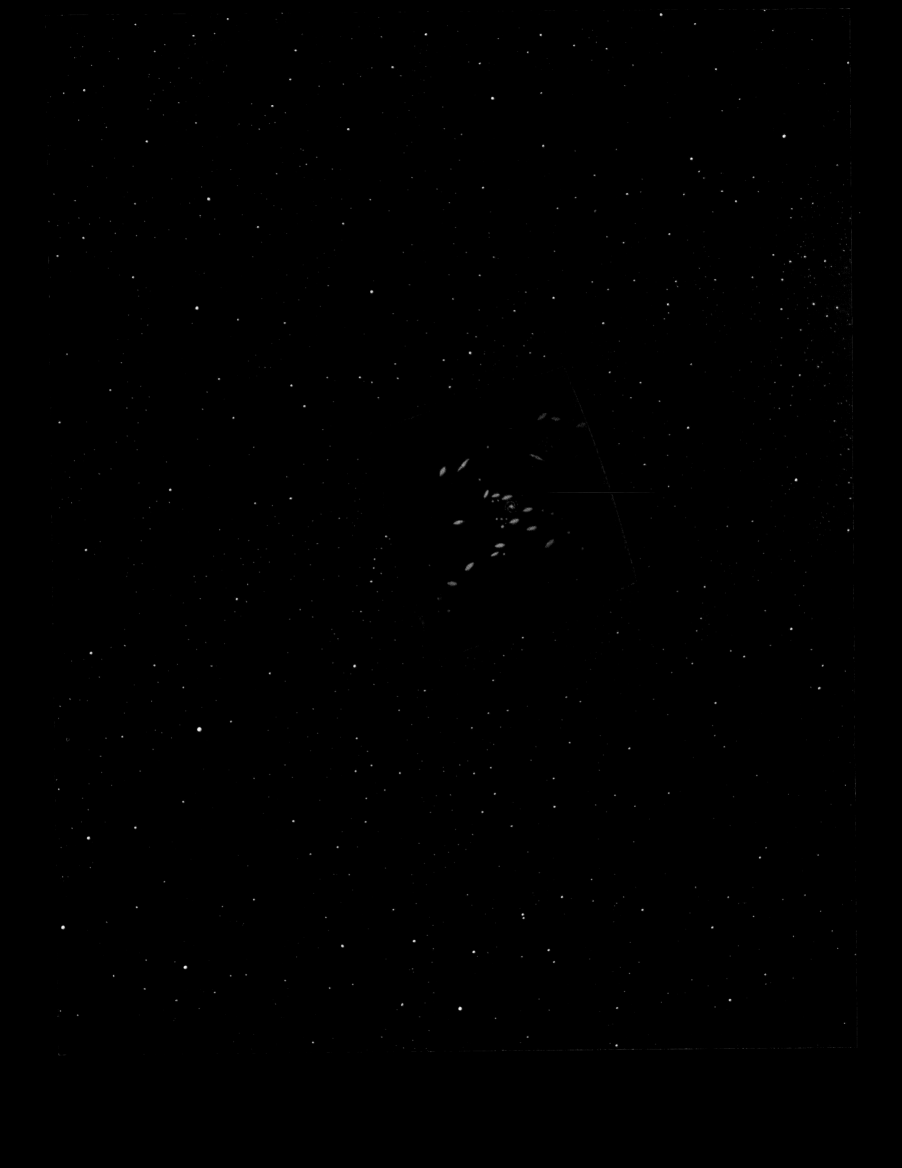

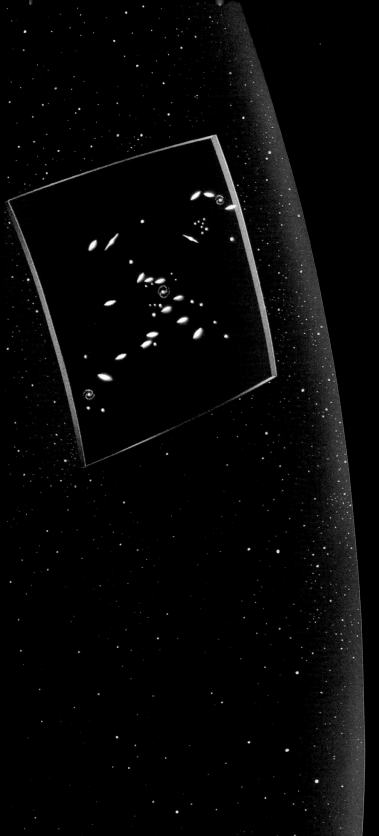

REDSHIFT RIDDLES

The first hint that concentrations of mass larger than individual galaxies might be exhibiting peculiar motions grew out of efforts to measure the small-scale, gravitationally induced movements of galaxies within a few tens of millions of light-years of Earth. Astronomers calculate a galaxy's velocity by using its light as a kind of celestial speedometer. The faster it is moving away from Earth, the more the constituent wavelengths of its light will be shifted toward the lower-frequency, red end of the spectrum—just as the pitch of a train whistle will drop as the train speeds away. Some of the velocity reflected in redshift is the product of Hubble flow. Because Hubble flow velocity is directly proportional to a celestial object's distance from Earth, astronomers can determine that velocity component by estimating distance and then subtracting the Hubble flow value from the observed velocity to yield the "peculiar velocity."

In the mid-1970s, researchers applying these methods made a startling discovery. The Milky Way and its surrounding cluster of galaxies, the Local Group (shown at left, again within the context of a balloon cosmos), were not only moving apart with the expansion of space but also moving as a group at some 400 miles per second. Such large-scale peculiar motion could not be due to nearby gravitational influences.

Mistrustful of their results, investigators sought a new backdrop against which to track the unexpected movement. They found it in the cosmic background radiation—the energy released shortly after the birth of the universe. Stretched to low-energy radio wavelengths by the expansion of space, the ubiquitous, once-intense radiation now reads a uniform 2.73 degrees Kelvin above absolute zero. Astronomers realized that if the Local Group of galaxies was in fact moving as indicated, the motion would have the effect of compressing the wavelengths of Big Bang radiation, causing a slight rise in the cosmic background frequency—a hot spot in the direction of their travel.

Southward on the great sphere of sky surrounding the Earth, balloon-borne instruments discovered the telltale signal: an increase of a few thousandths of a degree. Drawn by some as-yet-undetected mass, the Local Group was indeed rushing sidelong across the expanding fabric of space.

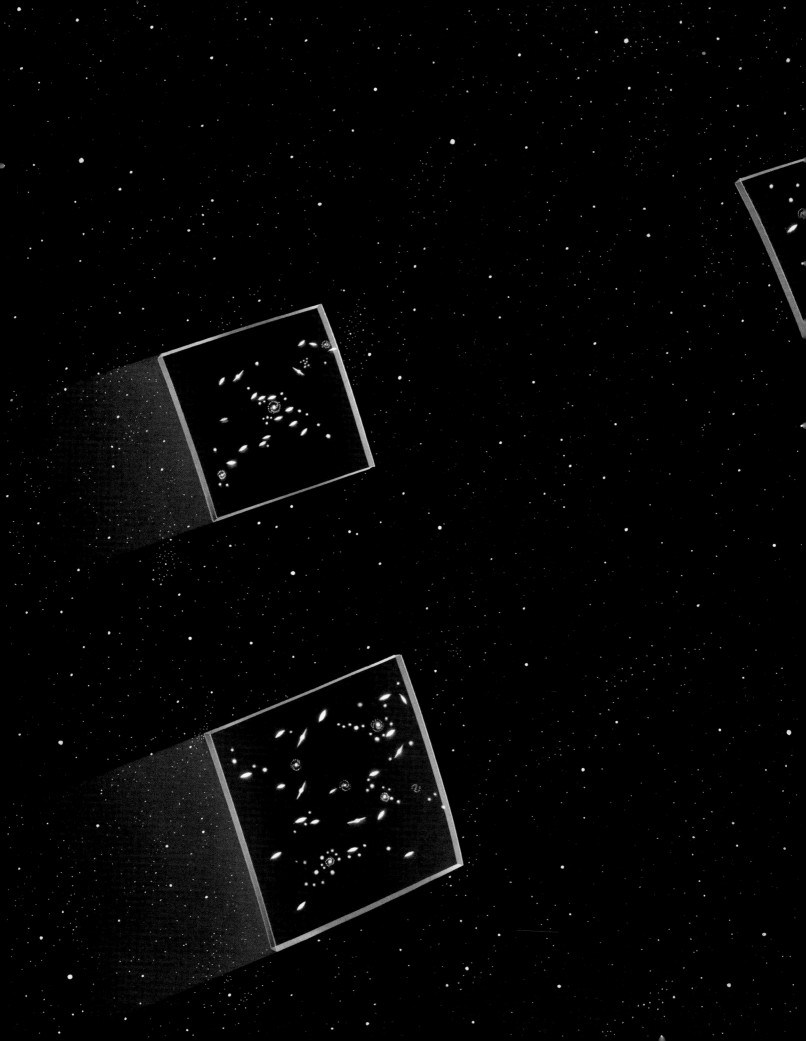

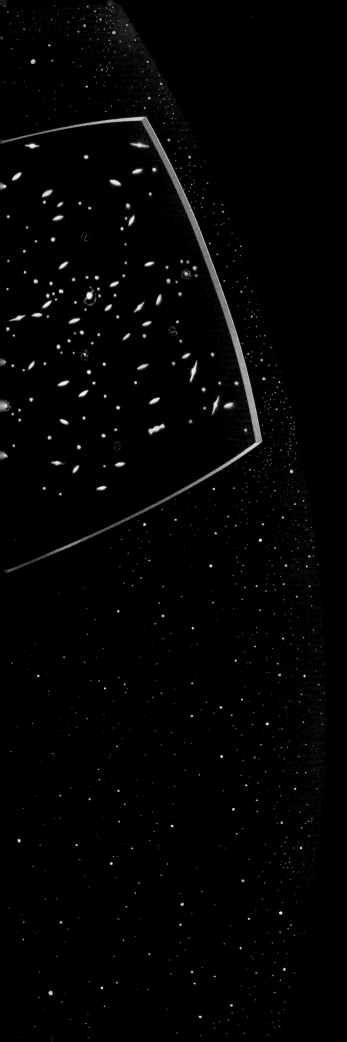

Mass Migrations

Unless the Local Group's mysterious motion was generated by shock waves forged during the explosive epoch of galaxy formation, something exceedingly large was pulling at the Milky Way and its starry entourage. Since any movements dating from the primordial era of star building should have long ago been lost in the overall Hubble flow, astronomers began searching for a gravitational behemoth—one capable of hauling a whole cluster of galaxies toward it.

In the simplest sense, tracking the source of the motion was a bit like following a trail of crumbs to a cookie jar. All galaxies within the sway of such an attractor must exhibit peculiar motions in its direction. The velocity of galaxies nearest the attracting mass will be the greatest, owing to gravity's stronger pull. The motions of those within the outer fringes of the mass itself will begin to slow, while galaxies at the very center, suspended in a perfect balance of gravitational forces, will show no peculiar motions at all.

Employing this reasoning, researchers mapped the peculiar velocities of several hundred galaxies within a 50-million-light-year radius of the Local Group. Gradually, like cosmic jigsaw pieces, the motions assumed a pattern. Outward from the Local Group *(far left, top)*, peculiar velocities increased steadily in the direction of the Virgo cluster *(far left, bottom)*, a group of galaxies at the heart of a vast swarm of some fifty galactic clusters termed the Local Supercluster. Galaxies on the opposite side of Virgo exhibited a delayed recession rate, indicating that they, too, were being pulled toward the Local Supercluster's center.

However impressive, though, the agglomeration of galaxies in the Virgo cluster could account for only part of the Local Group's peculiar motion toward the Local Supercluster. Moreover, the hot spot in the cosmic background radiation was focused in a region of space fifty degrees away from Virgo's center. When astronomers combined the motions toward Virgo and the hot spot in the background radiation, they discovered that most of the Local Group's motion was influenced by an even more distant lure: the massive Hydra-Centaurus supercluster, 100 million light-years away *(near left)*. And the Local Group was not journeying there alone; the entire Local Supercluster was streaming along.

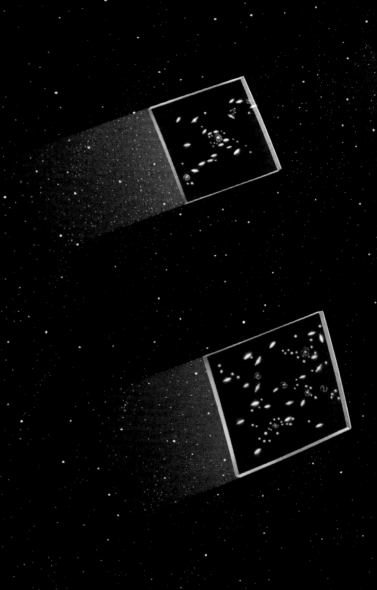

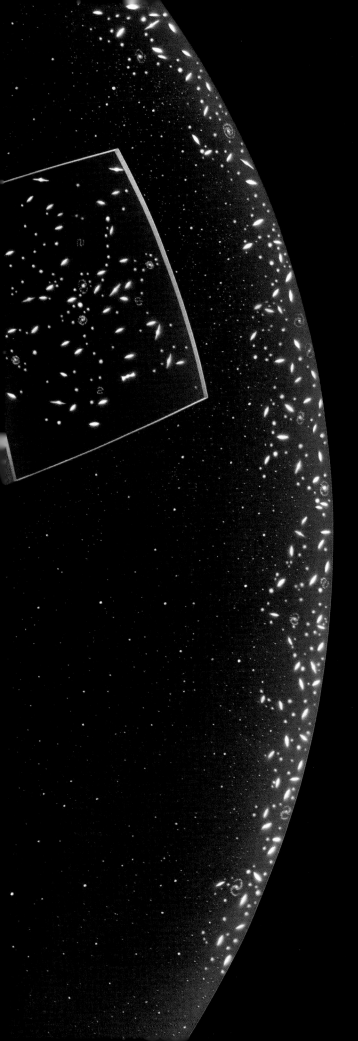

THE GREAT ATTRACTOR

Although the evidence was undeniable, scientists were baffled: To account for the Local Supercluster's movement *(far left, top)*, Hydra-Centaurus *(far left, bottom)* had to contain ten times the mass of the Virgo assemblage—far more mass than its luminosity indicates. Spurred by this cosmic paradox, researchers embarked on an unprecedented study of 400 elliptical galaxies sprinkled across a swath of sky 200 million light-years wide. Painstakingly, over a six-year period, they mapped the galaxies' peculiar motions.

Instead of providing answers, however, the study presented scientists with a stranger riddle still: Galaxies in Hydra-Centaurus, the purported siren, were themselves in motion. Under the influence of some monster mass, both Hydra-Centaurus and the Local Supercluster were observed to be streaming in concert through space at better than 400 miles per second over the cosmic expansion rate.

Dubbed the Great Attractor, the source of the mysterious peculiar motions became the subject of an intensive cosmological hunt. Investigators charted the peculiar motions of galaxies through increasing volumes of space surrounding Hydra-Centaurus. At about 100 million light-years from Earth, on the fringes of an immense collection of spiral galaxies *(near left)*, researchers detected the first decline in individual galactic movement. Peering 50 million light-years deeper, they encountered the Attractor's very heart: a region where all peculiar motion has ceased and galaxies are at rest with the Hubble flow. On the far side of the Attractor's center, the recession velocities of galaxies were slowed by gravity's inexorable grip, producing a so-called infall.

Based on its drawing power alone, astronomers have estimated the Attractor's mass at thirty times that of the Local Supercluster, the equivalent of 10,000 trillion Suns. Although the dusty plane of the Milky Way obscures a large portion of this spectacular galactic collection from Earth view, it is believed to be a flattened supercluster some 100 million light-years thick and twice as wide. The perpetrator at last in hand, astronomers now face the greatest challenge of all. They must reconcile their new vision of cosmic lumpiness with the prevailing view of how the universe was born. The two do not seem to jibe.

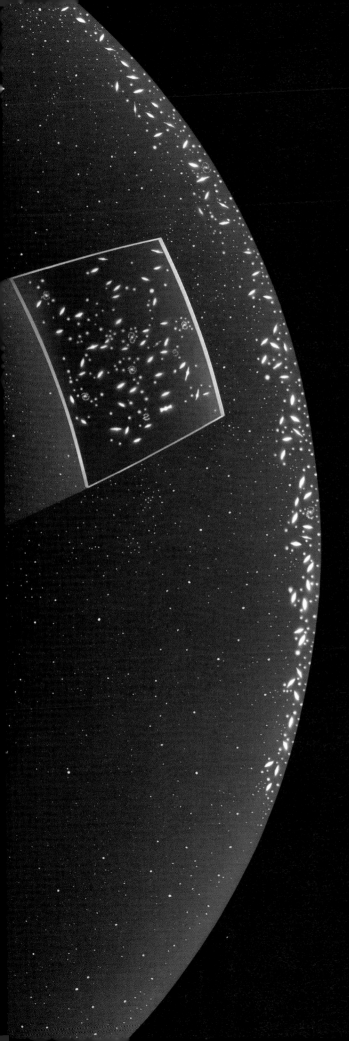

Cosmic Destiny

Routine galactic surveys performed during the 1980s had already revealed considerable unevenness in the large-scale structure of the universe: The cosmos appeared to be a web of bubblelike voids surrounded by galactic filaments, each filament consisting of hundreds or thousands of galaxies. These findings called into question the smoothness of the cosmic fabric, and the lumpiness evidenced by the Great Attractor is obliging astronomers to rethink the Big Bang altogether. For stupendous conglomerates such as the Great Attractor to have developed, large-density fluctuations had to be present in the matter and radiation soup of the early universe—a condition that should now be reflected in the background radiation. But no such irregularity has yet been found.

One explanation for the discrepancy hinges on cosmic strings—predicted (but unproven) remnants of the Big Bang. Cosmic strings are envisioned as massive concentrations of primal energy and gravity; if they exist, they might be capable of gathering matter into metasuperclusters. Another possibility is that matter is evenly distributed after all; instead of existing primarily in radiant form, much of it may be secreted in exotic dark particles. However, recent evidence suggests that dark matter may be as lumpy as its luminous counterpart. A gargantuan confluence of galaxies known as Perseus-Pisces-Pegasus *(far left)*, on the side of the Local Supercluster *(left, center)* opposite from the Great Attractor *(near left)*, does not appear to be drawing galaxies toward it. If, as some measurements indicate, both superclusters are equally bright, a dearth of dark matter in Perseus-Pisces-Pegasus seems the only explanation.

Astronomers hope that the large-scale peculiar motions of galaxies will provide a window on the fate of the universe. By mapping the streaming velocities of galaxies over greater regions of space, they might discover the distribution of gravity, and hence matter, in the cosmos. From this, they could estimate the universe's average mass density. A number less than a critical value set by the equations of general relativity will signify an "open" universe, destined to expand into oblivion. A higher number will portend a closed cosmos, in which galaxies turn in upon themselves, racing to destruction—and perhaps to rebirth.

3/Into the Unknown

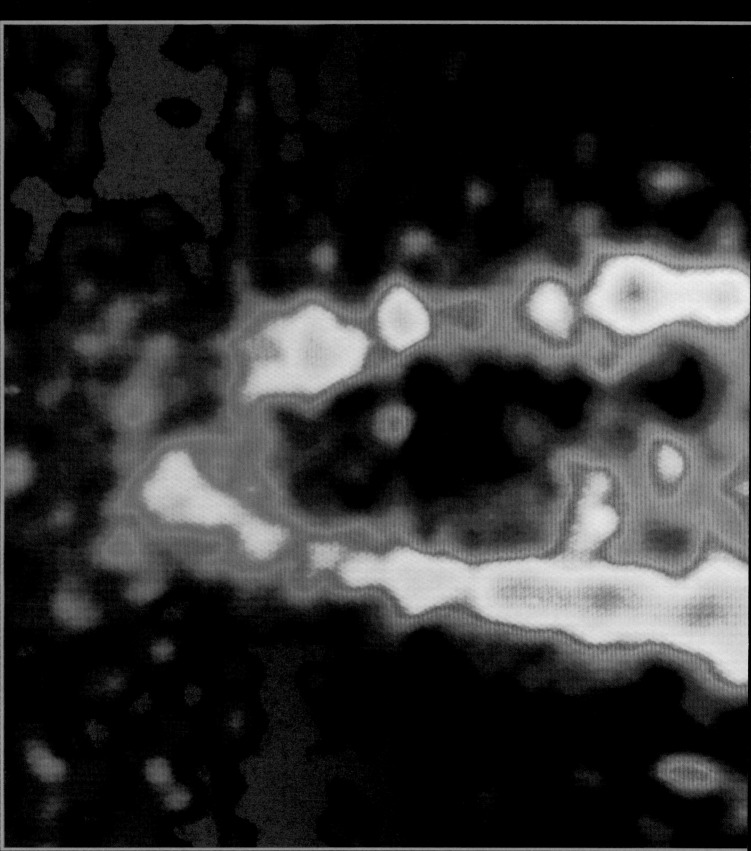

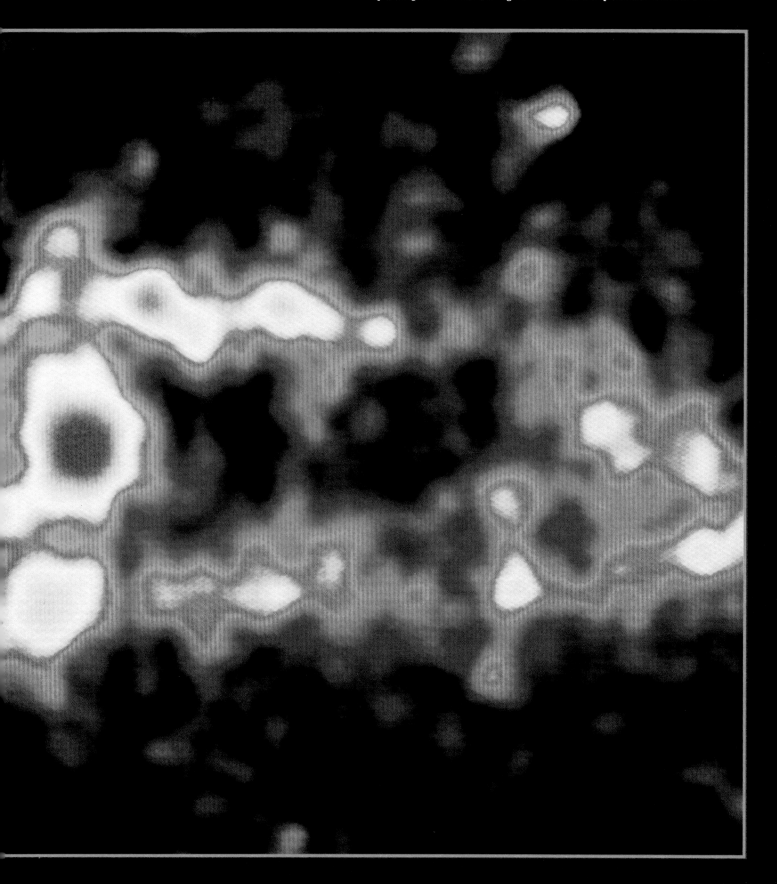

Andromeda's engine. On a radio map of the great spiral galaxy Andromeda, synchrotron radiation emitted by electrons spiraling around powerful magnetic lines of force increases in intensity *(orange)* at the core, a region brimming with stars. Along with revealing the presence of gas near the center traveling at 200,000 miles per hour, the radiation hints at a gravitational monster in the galaxy's heart—a likely candidate for a supermassive black hole.

During the twentieth century, revolutionary developments in theoretical physics have so transformed scientists' view of the cosmos that the classical Newtonian description now appears almost rudimentary. The universe as envisioned by astronomy's theorists teems with exotic phenomena, from the strange, mirror-image particles of antimatter to the bewildering gravitational realms of black holes, white holes, and wormholes—tunnels through the fabric of space-time. Astrophysicists even speculate that whole other universes are continually springing into existence. Many of the theorists' predictions lie beyond the bounds of detection by their very nature, to be pursued only as extreme solutions to mathematical equations.

Time and again, even when physical proof emerges, new questions arise, leading to increasingly abstruse conjectures. But no matter what queer paths the imaginations of astrophysicists take, the universe always seems to stay at least one step ahead, teasing the human intellect with ever-deeper secrets.

BREAKING NEW GROUND
Among the casualties of the revolution in physics was the accepted model of matter itself. The theory that matter is composed of atoms, although conceived by the Greek philosopher Democritus as long ago as the fifth century BC, was not confirmed until the 1890s. For a number of years, scientists' knowledge of this submicroscopic world remained sketchy. They assumed—naturally enough—that the particles of which atoms are composed followed the same laws that ruled the motions of heavenly bodies. Electrons, for example, were thought to orbit an atom's nucleus just as Earth and the other planets of the Solar System circle the Sun.

By the 1920s, however, the world of physics was in upheaval, the conventional wisdom having already suffered repeated challenges. Quantum theory, introduced by German physicist Max Planck at the turn of the century and later expanded by Albert Einstein in 1905, described energy as possessing a dual nature, sometimes behaving as a wave and sometimes as discrete packets, or quanta, known as photons. Several years later, Danish physicist Niels Bohr used this notion to refine the model of the atom, explaining that electron orbits represent distinct energy levels and that by absorbing or emitting a fixed amount of energy, an electron can jump from one orbit to another. Then, in 1924, French theoretician Louis de Broglie suggested that electrons themselves display a split personality and often act more like waves than particles.

The apparent paradox stirred fierce debate and quickly led to other bizarre concepts. German physicist Werner Heisenberg demonstrated that it was impossible to define an electron's momentum and its exact position at the same time. Rather than blaming the inadequacies of current detection methods, he and Bohr argued that this was an inherent characteristic of the subatomic world. Over the objections of no less a figure than Einstein, who refused to accept that God would play dice with the universe, the leading view became that an electron could be described only in terms of probabilities, not certainties. Among the most unsettling aspects of this idea was that an electron could vanish from one spot and reappear in another without traversing the space in between. (Today, such counterintuitive behavior is demonstrated in many practical applications: For example, the scanning tunneling electron microscope, which has produced extraordinary images of the surfaces of molecules, works because electrons spontaneously disappear from the tip of a needlelike probe and emerge on the surface of a scanned object, setting up an electrical current that traces the object's structure at the atomic level.)

MATTER'S MIRROR IMAGE

Into the scientific maelstrom of quantum theory came a young man named Paul Dirac. Born in Bristol, England, in 1902, Dirac was immersed in an academic atmosphere from an early age. His father, a Swiss immigrant, was the French teacher at the local school and encouraged his son to become bilingual by demanding that he use no English in his presence. An excellent student with a particular aptitude for mathematics, Dirac earned his undergraduate degree in engineering from the University of Bristol, then moved on to Cambridge for graduate work in physics and mathematics. He quickly proved to be among the most brilliant physicists of the day, making fundamental contributions in several areas of the new quantum physics. His most celebrated work was a set of equations published in 1928 that described the behavior and properties of the electron in detail. There was only one problem with his equations: They predicted the existence of a new particle, identical to the electron in every respect but with a positive electrical charge.

The idea was roundly criticized by Dirac's colleagues. At the time, the scientific community was in almost universal agreement that only three elementary particles existed: the electron and the much more massive proton and neutron, which together formed the atomic nucleus. There was absolutely no corroborating evidence that the subatomic world was any more diverse than this, nor was there any reason that it should be, beyond Dirac's somewhat arcane hypothesis.

Four years later, however, an experiment proved Dirac's claim to be true. The man responsible was Carl David Anderson, an American physicist working at the California Institute of Technology. Ironically, Anderson was not looking for Dirac's antielectron at all. His specialty was cosmic rays, atomic fragments that rain down on Earth at extremely high speeds, possibly from such sources as solar flares or supernovae. Various detection methods had revealed that as these bits of cosmic matter enter Earth's atmosphere, they collide with molecules in the air and shatter them, releasing bursts of energy in the form of gamma rays and creating secondary showers of subatomic particles. Both the gamma rays and the secondary particles, also highly energized because of the violence of the collisions, go on to generate a continuing cascade of fragments, some of which reach Earth's surface.

To study the process more closely, Anderson employed a device known as a cloud chamber—a sealed container filled with moist air, originally designed to simulate the formation of clouds in the upper atmosphere. The air molecules in the chamber would provide targets for incoming cosmic rays and ensure the generation of subsequent particle showers; a lead plate dividing the box horizontally would slow down particles enough to permit accurate measurements of their properties. As they passed through the chamber, these subatomic fragments would cause the water vapor to condense into tiny droplets along the particles' path, thus rendering their passage visible. A strong magnetic field applied to the chamber would also cause particles of opposite charge to act in different ways, enabling Anderson to distinguish the kinds of particles being produced.

Once the experiment got under way, he noticed the presence of a certain type of particle with all the expected properties of an electron but curving in the opposite direction, a clear indication that it carried a positive rather than a negative charge. Anderson's discovery, which he dubbed the positron, was the particle Dirac had predicted.

The prediction and the discovery led to Nobel prizes for Dirac and Anderson and ignited a new flurry of theoretical and observational work. Further cosmic-ray studies, for example, revealed the existence of the muon, a negatively charged particle similar to the electron but 200 times more massive and highly unstable, decaying within millionths of a second into still other kinds of particles. It soon became apparent that the basic constituents of matter were far more various than anyone had imagined. But perhaps even more fascinating was the realization—confirmed by a series of experiments during the 1950s and 1960s—that the electron-positron relationship is standard in the subatomic world: For each type of matter particle there is an antimatter equivalent that is opposite in electrical charge or some other fundamental property.

Researchers began to learn more about the antimatter realm by using particle accelerators, which boost subatomic particles to tremendous speeds and then smash them apart, mimicking the effects of cosmic rays and yielding antiparticles by the score. Also, bubble chambers—sophisticated versions of the cloud chamber filled with liquid hydrogen rather than water vapor—have permitted more detailed studies. Scientists now know, for instance, that when gamma rays collide with atomic nuclei, their energy is sometimes

Symmetry and asymmetry. When high-energy gamma rays are fired into a bubble chamber filled with liquid hydrogen, collisions with hydrogen atoms can produce electrons (shown in green in the two collisions at right) and their antimatter counterparts, positrons (red). In some collisions (top), an electron knocked loose from the atom carries off much of the energy of the interaction, causing the electron-positron pair to spiral in tight arcs; the more energetic the particles, the straighter the trails (bottom pair). A magnetic field in the bubble chamber causes the oppositely charged particles to follow mirror-image curved paths. Although the symmetrical creation of matter and antimatter is common in such experiments, the universe outside the physics laboratory is dominated by matter—an asymmetry cosmologists find baffling.

converted into both positrons and electrons. Furthermore, controlled experiments have confirmed another of Dirac's original predictions: Whenever one or more particles and antiparticles meet, they annihilate each other in a burst of pure energy.

A DEARTH OF ANTIMATTER
As physicists improved their understanding of how antimatter forms and how it behaves, a puzzling discrepancy emerged. Anderson had observed that cosmic rays produce as many positrons as electrons, and later findings indicated that gamma ray collisions can result in matched pairs of particles and antiparticles. In addition, although a positron and an electron curve in opposite directions under the influence of a magnetic field, the patterns are exact mirror images of each other, suggesting that matter and antimatter behave in symmetrical fashion with respect to the laws of physics. The implication was obvious: Extremely energetic processes that create matter should just as easily create antimatter. One such process, of course, was the formation of the universe, in which matter and energy came into being. Given the dynamics of the forces at work shortly after the Big Bang, antimatter should be just as abundant in the cosmos as matter.

Where, then, is it? Clearly, no antimatter exists in any appreciable amount on Earth; if it did, it would readily come into contact with matter and vaporize in huge explosions. And since Earth is made of matter, the Solar System must be also, because the Sun and all the planets condensed out of the same cloud of dust and gas. As for the entire galaxy, if there are such things as antimatter stars, some would already have gone supernova, pouring vast quantities of antiparticles into the interstellar medium and thereby producing almost constant matter-antimatter annihilations and their telltale bursts of energy. Similar arguments hold for other galaxies: If anywhere near an equal number of them are made of antimatter, galactic collisions would cause incredibly powerful blasts that could easily be detected from Earth. The only possibility for the presence of significant quantities of antimatter is that it resides in whole clusters of galaxies separated by millions of light-years from the matter-dominated galaxies in Earth's corner of the cosmos—a highly doubtful proposition, since there is no way to explain what could have propelled such a dramatic separation. With few exceptions, astronomers thus agree that the universe is virtually all matter.

Based on what cosmologists believe about the universe's earliest stages, the real question is not why antimatter is so scarce but why matter exists at all. According to standard theory, particles—and, presumably, antiparticles—began to form an infinitesimal fraction of a second after the Big Bang, as the cosmos started to expand and cool. The matter and antimatter would have been close enough together to annihilate each other almost instantaneously, and although they would continue to come into being in what was still an extremely hot universe, the annihilations would have kept pace. If the amounts of matter and antimatter had been exactly equal, there eventually would have been nothing left but energy—nothing from which stars and galaxies could have formed.

The solution to this paradox is far from certain, but scientists now believe it is rooted in a discovery made in 1963 by Princeton physicists James Cronin and Val Fitch. Through a complex chain of logic based on experiments they performed at the Brookhaven National Laboratory in New York, Cronin and Fitch proved that, contrary to expectations, matter and antimatter do not respond identically or symmetrically to all of the four forces in nature—gravity, electromagnetism, and the strong and weak nuclear forces. They found that the weak nuclear force, which governs radioactive decay, made a crucial distinction between matter and antimatter in regard to the breakdown of a particle known as the neutral K meson.

Despite more than two decades of additional research, the reason for this differentiation is still not apparent, nor is there any direct evidence that it would lead to a dominance of matter over antimatter; in fact, the experiments conducted by Cronin and Fitch actually showed neutral K mesons decaying into positrons more often than electrons. For cosmologists, however, the significant point was the demonstration of unevenness in the behavior of a natural force. In the first moments of the universe, such an asymmetry could have had the effect of allowing an initially precise balance between matter and antimatter to evolve quickly into the slightest of imbalances. That imbalance need only have been a billion and one particles of matter for every billion particles of antimatter. Even after an extended orgy of mutual destruction, there would still have been enough matter left over to account for the current state of the universe.

ON THE BLACK HOLE TRAIL

The theory of antimatter developed relatively quickly from its origins in quantum physics and was confirmed by observation in short order as well. In the case of black holes, however, the story has unfolded much more slowly—over the course of centuries rather than just a handful of years—and has yet to reach a definitive conclusion.

The earliest recorded mention of what is now called a black hole came in a presentation to Britain's prestigious Royal Society on November 27, 1783, by a man named John Michell, a professor of geology at Cambridge. Michell is remembered today for important discoveries in the study of earthquakes, but his innate scientific curiosity led him to explore a number of subjects, including the nature and effects of gravity, which had been mathematically described by Isaac Newton less than a hundred years earlier. In particular, Michell focused on the notion of escape velocity: In order to break free from the gravitational field of a planet or a star, an object has to be moving with a certain minimum speed, or else it will fall back. The actual speed depends on the mass of the celestial body and the location of the object within the body's gravitational field. For example, Earth's escape velocity, measured from the planet's surface, is 7 miles per second, while the Moon's is 1.5 miles per second. What Michell did was to postulate a celestial body so compact that its escape velocity would exceed the speed of light—estimated at the time to be about 183,000 miles per second, only some 3,000 miles per second shy of the actual value. He concluded that "all light emitted from such a body would be made to return to it by its own proper gravity."

Michell's concept was essentially ignored by the scientific community, perhaps because it was outside his field of expertise. Thirteen years later, French astronomer Pierre-Simon Laplace independently reached the same result in a work on celestial mechanics, predicting that an extremely massive object—its diameter 250 times that of the Sun but as dense as Earth throughout—would be so gravitationally powerful that light could not escape it. Laplace deserves credit for introducing the idea to the mainstream of astronomical thinking, even if Michell had actually beaten him to the punch. Nonetheless, the scientific community did little more than admire Laplace's cleverness in devising such an interesting exercise in Newtonian physics; no one at the time believed such a body could really exist.

Michell's and Laplace's theories, although long overlooked, are still among the most accessible descriptions of black holes, and their estimates of the necessary mass have turned out to be surprisingly accurate. However, a

A Lineup of Black Hole Theorists

1796 Pierre-Simon Laplace, believing that light, like ordinary matter, is slowed by gravity, postulated that very large stars would prevent light from escaping and would thus be dark.

1915 Albert Einstein's general theory of relativity, introducing curved space-time, became a tool for understanding the bizarre gravitational effects of dense stars and black holes.

1916 Karl Schwarzschild solved Einstein's equations to work out the distance—the Schwarzschild radius—within which a given mass would be so dense that its gravity would trap light.

1939 Robert Oppenheimer described how a large star that has spent its thermonuclear fuel would collapse to a dimensionless point, or singularity, inside the Schwarzschild radius.

more refined and ultimately more precise understanding of how gravity affects light depends on the entirely new way of thinking about gravitation that was conceived by Einstein early in the twentieth century.

RELATIVITY TO THE RESCUE

Einstein's special and general theories of relativity established him as one of the greatest scientists of all time. The former, which was first published in 1905, introduced the unsettling notion that measurements of time and distance vary, depending on the relative motions of an observer and an object being observed; the velocity of light, however, is an unvarying constant. From these premises emerged a complete reformulation of the concepts of space and time. Both entities are actually aspects of a single reality called space-time, a cosmic stage of four dimensions (three spatial and one temporal) on which the interactions of matter and energy are played out. Although the idea is difficult to grasp, it becomes somewhat more comprehensible when one realizes that light always takes a certain amount of time to travel from place to place; therefore, any object viewed at a distance—even at arm's length—is removed not only in space but also in time. (Of course, light is so speedy that the effect is vanishingly small except on a celestial scale.)

Einstein's next task, accomplished in the general theory of relativity, was to work gravity into the picture. The Newtonian description of gravity as a force fails to explain how objects pull on each other when there is nothing connecting them. A better approach, Einstein argued, was to imagine gravity

1963 Roy Kerr found a new solution to Einstein's equations: a ring-shaped singularity with spin as well as mass, surrounded by an "ergosphere," from which objects might escape.

1969 Roger Penrose proposed extracting power from a Kerr black hole via its ergosphere: An escaping particle would gain energy taken from the hole's mass and angular momentum.

1971 Stephen Hawking conceived of low-mass singularities, spawned not by collapsing stars but by the violent turbulence of the Big Bang. Such mini black holes may still exist.

as distorting the fabric of space-time. Since four dimensions are impossible to visualize, the easiest way to understand this is to think of space-time as a two-dimensional rubber sheet in which mass creates dips, known as gravity wells; the more massive—and thus the more gravitationally powerful—an object, the deeper the well. In this scenario, a ball thrown into the air falls back to Earth not because of some mystical attractive force but because it does not have enough energy to climb up and over the sides of the well. While the distortions in Einstein's space-time are far more subtle than dimples in a rubber sheet, the analogy is a close one and has become a standard among astrophysicists. Furthermore, in the decades since the theory was proposed, it has shown itself to be unfailingly accurate at describing and predicting the behavior of heavenly bodies.

As it happens, the mathematical formulations of general relativity were so incredibly complex that Einstein himself had settled for approximate solutions as to how matter affected the structure of space-time. However, almost immediately upon the 1915 publication of the theory, a German astronomer and theoretical physicist named Karl Schwarzschild took up the chore of producing exact results; by May of 1916, he had solved the equations for the curvature of space-time around any spherical mass. The feat was remarkable, especially considering that Schwarzschild, who had volunteered for the army early in World War I, was serving on the Russian front when he first began working on the problem and had also contracted pemphigus, a rare viral skin disease that would kill him shortly after he completed his calculations. Included in the solutions was the specific—and, as far as anyone knew, purely theoretical—case where all the mass is concentrated with infinite density at a single point, a phenomenon known as a singularity. The result is a space-time curvature so pronounced that anything trying to escape, including a ray of light, would be unable to do so, bending back into the singularity instead.

According to Schwarzschild's calculations, while the singularity itself is dimensionless, its light-trapping gravitational effect extends to a defined volume of space surrounding it: Escape is impossible from anywhere in that space. Together, the two features constitute what scientists now call a black hole. (The term was coined by physicist John Wheeler of Princeton in 1967.) The encompassing region's limit, known as the Schwarzschild radius, depends on the amount of mass within the singularity. For a black hole with the Sun's mass, for example, the radius would reach about two miles. Physicists also refer to this boundary as the event horizon: Since light is entrapped, any event that takes place inside is "over the horizon," or beyond the view of an observer.

Einstein was impressed by Schwarzschild's mathematical skill, but he was extremely uncomfortable with the idea of infinite density, primarily because it suggested that a finite, measurable quantity of mass could occupy a place that was, literally, measureless. At the time, however, since there was no known mechanism in the physical world capable of squeezing matter into

In a one-page parody of the *New York Times,* theoretician Kenneth Brecher heralded a 1974 physics colloquium at the Massachusetts Institute of Technology with spoofs of alleged evidence for black holes, facetious treatments of recent discoveries, and the "first detailed color photograph of a black hole": the black square at lower right.

The Black Hole

"No News Escapes Us"

Abandon Hope All Ye Who Enter Here

Thursday, February 14, 1974

Announcement of M.I.T. Physics Colloquium By K. Brecher

Les Corps Obscurs de Laplace-Existent-Ils?

PARIS, 1796 — At a recent meeting of L'Academie des Sciences, M. Le Marquis De Laplace, the eminent mathematician and natural philosopher, provided for all those present a most amusing and entertaining evening. With readings from his recent best seller "Exposition Du Systeme Du Monde," while circulating amongst the audience reprints of his latest paper in the Allgemeine Geographische Ephemeriden, he presented a talk entitled "Future Progress of Astronomy." Amongst other speculations, he suggested that the Universe is filled with "des corps obscurs," dark bodies, in numbers equal to the visible stars! He bases these ideas on his calculations which show that "a luminous star, of the same density as the earth, and whose diameter should be 250 times larger than the sun would not, in consequence of its attraction, allow any of its rays to arrive at us." He concluded by saying that "it is therefore possible that the largest luminous bodies in the universe may, through this cause, be invisible." Despite the irrefutability of his mathematics, he failed to suggest how any object would come to exist in such an ignominious state. One can only hope that his good name will not be darkened by such flights of fantasy. (Ed. note — By the publication of the fifth edition of "The System of the World" Laplace had expunged all references to "des corps obscurs.")

Book Review:
"The Other Side"
by Alfred Kubin

VIENNA, 1909 — In a fit of brilliant insight and intense productivity, the great Austrian presurrealist painter Alfred Kubin has succeeded, where no man has before him, in grasping the full physical significance of collapse into a black hole. A brief illustration from his novel "Die Andere Seite" should suffice to support this claim. Turning from the brush to the pen, he wrote: "And now, for the first time, I discovered in the veil of mist an immense, high wall. Suddenly, unexpectedly, it loomed up before me. Someone carrying a light was walking in front of us toward an enormous black hole: that was the gate to the Dream Kingdom. As we approached I noticed its huge dimensions. We entered a tunnel, keeping as close as we could to our guide. Then something strange happened. I had already penetrated some distance into the vaulted passage when I was overcome, as though at a blow, by a wholly unfamiliar and dreadful sensation. It began at the back of my head and ran down my spine; my breath stopped, and my heart beat wildly. Helplessly I looked toward my wife, but she herself was white as a corpse, deathly fear mirrored in her face. In a quivering voice, she whispered: 'I shall never come out of here again.'" His recognition of the role of tidal forces and of the irreversability of such a predicament are all the more remarkable for they predate Herr Einstein's General Theory of Relativity by seven years.

SCIENTISTS FORESEE: COLLAPSE INEVITABLE

BERKELEY, 1939 — Out of the depths of the Great Depression, and confronted with the possibility of another worldwide conflagration, the brilliant young American physicist J. Robert Oppenheimer and his graduate student, former truck driver Hartland Snyder, have reported in the latest issue of the Physical Review that "when all thermonuclear sources of energy are exhausted, a sufficiently heavy star will collapse." Such news should be kept in mind by those who would hope that a detente could be achieved by bringing pressure to bear on arbitrarily large bodies to counter the ever present gravity of the situation. Furthermore, as the authors are the first to point out, while a sufficiently distant observer will never see its final demise, a person collapsing with a massive body will experience all the accompanying stresses in less than a day.

CYGNUS X-1: BLACK HOLE OR RED HERRING?

Popular model of Cyg X-1, consisting of a binary star system containing a black hole (at the center of the disk, lower left) accreting matter ejected from its more massive companion.

Exciting Young Star Finds Happiness With Old Degenerate Dwarf

BOSTON, 1973 — On a day with very little news reaching us, a hopeful and touching story has emerged. It is commonly believed that overweight old stars have no alternative but to eventually collapse and disappear from sight altogether. Not so, say two MIT Professors, K. Brecher and P. Morrison. In a surprising twist of the usual scenario, they suggest that such stars can avoid this fate by turning instead into degenerate dwarfs. If they get around enough, such stars can again become radiant and even, as they suggest in the case of Cygnus (The Swan) X-1, co-habitate with a star as young and bright as HD226868. (Ed. note — This story should satisfy those readers who have accused us of a discriminatory publishing policy. It is only the first in our new affirmative action series featuring such recently neglected stars as white dwarfs, red giants and, if space permits, blue stragglers. This series will complement our ongoing reports on the activities of some prominent white holes. Owing to cosmic censorship, however, we have been unable to uncover any information surrounding naked singularities.)

Princeton Professor Proclaims Black Holes Have No Hair

PRINCETON, 1972 — Professor John Wheeler of the Princeton University Physics Department, reporting on his own researches, as well as those of Drs. Penrose, Hawking and others of Great Britain, has revealed that should black holes be discovered soon, there is little to distinguish one from the other. This follows, he says, from very general and powerful mathematical theorems which imply that such a body is completely characterized by three independent quantities: mass, charge and angular momentum (see figure). Such a conclusion, however, may be premature as has been emphasized by Professor F. Curtis Michel in his recent article in the journal Comments on Astrophysics and Space Physics entitled "Hair Tonic For Black Holes." He cautions the unwary, "If black holes indeed have no hair, it could be because they have no scalp for it to grow out of. However, there is a lot of stuff floating around looking suspiciously like dandruff."

NEW YORK, April 1, 1971 — The New York Times today reported for the first time the discovery of a "black hole in space." Variously referred to as a "collapsar" (A.G.W. Cameron of the (Veritas) Center for Astrophysics) or "frozen star" (Ya.B. Zeldovitch of the Soviet Academy of Sciences), such objects have long filled the void of theoretical astrophysicists waking hours. Now at last, it seems, there is an object upon which they can lavish their speculations. Scientists from American Science and Engineering, Inc., headed by Dr. Riccardo Giacconi, making observations with instruments aboard the first small astronomical satellite, nicknamed UHURU, claim to have finally shed some light on the matter of black holes or, more precisely, say that they have seen the light, from matter spiralling headlong into the oblivion of a black hole. They interpret the x-ray emissions from Cyg X-1 as arising from gas flows in a close binary star system containing a massive young star ejecting unwanted matter, which then accretes onto its fully collapsed companion (see picture). Waving aside the objections of a dissident minority of scientists who question whether Cyg X-1 is fully collapsed, or massive, or accreting, or, even, whether it is in a binary star system, Dr. Giacconi told this reporter in no uncertain terms that ". . . uncertain (Continued on page 13)

Texas Teachers Tout Tunguska Tragedy

AUSTIN, 1973 — Waving aside as extravagant and speculative the claims by Russian scientists that the immense explosive event which occurred in the Tunguska region of Siberia on June 30, 1908 was a great meteorite or comet, two scientists at the Center for Relativity Theory at the University of Texas, A.A. Jackson IV and M.P. Ryan Jr., have explained the event as having resulted from the passage of a mini black hole through the earth. Their suggested test of the theory, by hunting through old ships' logs for any record of the expected air and sea shock disturbances accompanying the re-emergence of the black hole in the North Atlantic has so far been stymied by Russian refusals to provide the vital records. (Tass, the Soviet News Agency, comments: Bourgeois captialist Americans, in an attempt to discredit the greatness of the People's Meteorite, which fell within Mother Russia in 1908, have put forward the ludicrous suggestion that it was a black hole, that most degenerate of all western inventions . . .)

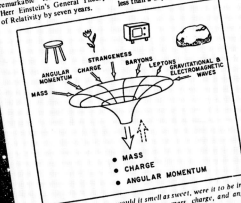

First detailed color photograph of a black hole. Note features at upper left and center, in good agreement with current theoretical predictions.

A rose is not a rose, nor would it smell as sweet, were it to be inside a black hole whose only attributes are mass, charge, and angular momentum.

such a state, Schwarzschild's concept, like Michell's and Laplace's, would remain a theoretical curiosity.

The properties of such an object were fascinating to contemplate, even if they had no counterpart in reality. Perhaps the strangest implication of Schwarzschild's work is the effect a black hole would have on time. Sometimes overlooked in the simplistic analogy of space-time as a two-dimensional sheet is the fact that time is one of the dimensions being distorted by gravity. According to the equations of general relativity, the stronger the gravitational force—or the deeper one's position inside a gravity well—the slower the passage of time. Thus, if a spaceship were to head toward a black hole and somehow avoid being torn apart by the tremendous tidal forces created by the hole's gravity, an outside observer would see something very odd happening. Although the ship would be picking up speed as it neared the Schwarzschild radius, to the observer it would appear to slow down; at the Schwarzschild radius, it would actually seem to stop, time coming to a standstill at the point where the escape velocity equaled the speed of light. Furthermore, in an effect equivalent to the Doppler shifting of receding objects, the spaceship would grow redder on its approach, as the wavelengths of light emanating from it were stretched out across an increasingly curved region of space-time.

As other theorists began to examine the notion of a singularity, it soon became clear that Schwarzschild's black hole was not the only possible kind. His solution indicated that the matter contained within a singularity would lose all its original attributes except mass; there would be no way of knowing, for instance, what its chemical composition had been. But the amount of mass could still be determined by measuring the strength of the gravitational field surrounding the singularity. Between 1916 and 1918, German physicist Hein-

A question of identity. Although it is only one star among thousands in the optical photograph above, a supergiant blue star named HDE 226868 *(red arrow)* has the distinction of being paired with a strong candidate for a black hole—an x-ray source named Cygnus X-1 *(right)*. X-ray sources within the galaxy are often the product of binary star systems in which x-rays are generated by hot gas spiraling from a supergiant onto a very small, very dense companion, usually a neutron star. But in calculating the size of this supergiant's unseen partner, scientists found that its mass must be at least five times that of the Sun—well beyond the theoretical limits for neutron stars. In theory, an object so small and yet so massive must be a black hole.

rich Reissner and his Finnish colleague Gunnar Nordstrom proved that if the matter had carried a net electrical charge before it reached infinite density, this quality also would be retained, since it would generate an electromagnetic field that would extend outward beyond the event horizon and thus could be measured.

This Reissner-Nordstrom singularity seemed even more improbable than the Schwarzschild black hole. Astrophysicists knew that large bodies such as Earth and the Sun are electrically neutral overall—a result of the fact that electrical opposites attract each other. Had Earth somehow begun as a negatively charged object, it quickly would have attracted positively charged particles from space until it evened out and became neutral. A black hole would presumably act the same way.

FROM POSSIBLE TO PLAUSIBLE

While the pursuit of the intriguing mathematical features of singularities continued, other theorists developed an understanding of stellar evolution that might lead to black holes. In 1932, Cambridge physicist Subrahmanyan Chandrasekhar demonstrated that the eventual fate of a dying star depends on its initial mass. Any star up to 1.44 times as massive as the Sun would collapse to the white dwarf stage but no farther; quantum mechanics indicated that for such a star, degeneracy pressure—in simple terms, the resistance of electrons to being squeezed together—would arrest the collapse. But what of larger stars? According to his calculations, Chandrasekhar knew that stars with more mass would overcome degeneracy pressure and collapse even farther, but the ultimate fate of such stars remained to be determined.

One possibility emerged from an unrelated hypothesis developed in 1934 by Walter Baade of the Mount Wilson Observatory and Caltech's Fritz Zwicky. Calculating that the degeneracy pressure of neutrons was even greater than that of electrons, Baade and Zwicky postulated the existence of neutron stars, which would be able to sustain even more mass than white dwarfs. But nothing in any of the equations worked out by Chandrasekhar and others indicated that this was the end of the story. A final answer came from, among others, J. Robert Oppenheimer, the brilliant Caltech physicist who would head the Manhattan Project, which developed the atomic bomb during World

War II. By incorporating all the effects of general relativity, Oppenheimer showed in 1939 that for any star more than about three times as massive as the Sun, even supposedly uncrushable neutrons would give way, and no force in the universe could keep the star from collapsing indefinitely. It would, he concluded, become a singularity.

Physics now had a mechanism to explain how black holes could actually come into being, not just descriptions of what form they would take after the fact. And it was now hard to imagine that the universe was not full of them; after all, stars of more than three solar masses were common throughout the galaxy, and at least some of them must have long since run out of nuclear fuel and died. A black hole was the only conceivable corpse they could have left behind. Yet there was little motivation to look for such an entity: How could something that was by definition invisible ever be seen?

Nonetheless, a new generation of theoretical physicists applied their talents to developing more elaborate descriptions of these odd objects. One major breakthrough came in 1963, when Roy Kerr, a mathematician from New Zealand, calculated the properties of a black hole that was not stationary but spinning. In retrospect, it was an obvious step: If black holes did form through the collapse of dying stars, they would naturally preserve the stars' angular momentum and actually pick up speed, just as does a shrinking neutron star. The result was a series of truly bizarre characteristics. For one thing, a Kerr black hole would have a more complex singularity at its core, shaped like a ring instead of a single point but still infinitely dense. As fantastic as it sounds, such a singularity would cause any object passing through the area inside the ring to emerge on the other side in a universe where gravity is a repulsive, rather than an attractive, force. Furthermore, a spinning black hole has what is known as an ergosphere, a region just outside the event horizon where space-time, although not curved in upon itself, is dragged around the hole, much like water in the vicinity of a drain. Escape is still possible from within the ergosphere, but only for objects moving at near light speed.

This so-called frame dragging led Oxford University mathematician Roger Penrose to another mind-boggling possibility. An advanced civilization, he suggested, could actually extract energy from the ergosphere. The idea sounds simple enough, although it is based on extremely complex principles. If a chunk of matter was injected into the ergosphere at a glancing angle so that it broke into two pieces inside, and if one of the pieces was to continue into the black hole while the other escaped, the latter piece would come out with more energy than the original matter had possessed. The extra power would have been stolen from the rotational energy of the black hole, and repeating the process would eventually cause the black hole to stop spinning.

EXPLODING BLACK HOLES

Next on the scene to explore the features of black holes was Cambridge physicist Stephen Hawking. Although he suffers from the degenerative illness known as Lou Gehrig's disease, he has for many years made great contributions to the understanding of the universe's mechanics. In 1974, he set about applying the laws of quantum mechanics to the region around black holes and came up with a shocking result: Contrary to everyone's assumption, black holes could actually radiate. Eventually, in fact, some of them might reach a point where they explode, spewing matter and energy back into the cosmos. As Hawking would put it years later in his popular book *A Brief History of Time,* "Black holes ain't so black."

This unusual proposition is based on the quantum-mechanical principle that matter is free to appear out of nowhere: Particles and their antimatter equivalents form spontaneously in supposedly empty space, then immediately annihilate each other—so quickly that their brief emergence into physical reality is by definition unobservable. Physicists know this happens because it subtly distorts the way atoms radiate energy in laboratory experiments.

Since pairs of these virtual particles (so named to distinguish them from actual ones) form everywhere in space, some must also spring up just outside the event horizons of black holes. Occasionally, Hawking reasoned, one member of a virtual pair will cross the event horizon before it can recombine with its partner, leaving the partner behind and thereby adding a new real particle (or antiparticle) to the universe.

The energy responsible for the generation of this particle can come only from the black hole itself, which is thus also the particle's source of mass. This means that as the particle escapes into space, the black hole is losing mass and energy: It is, in essence, radiating. (The captured particle does not add to the black hole's mass or energy, because it was originally a virtual component of the black hole's gravitational field.)

What are the ultimate consequences of this radiation? For black holes that formed from stars, the effect is negligible, since it would take much longer than the age of the universe for the black hole

Double black holes?
Outshining their host galaxies, two quasars *(yellow)* nearly touch in this false-color optical image of a rare binary quasar system 10 to 12 billion light-years away toward the constellation Pisces. The two galaxies may be exchanging material that could fuel the supermassive black holes believed to lie at the hearts of quasars.

to lose a substantial fraction of its mass and become unstable. But Hawking postulated that stellar collapse is not the only means by which black holes could form and that much smaller black holes might exist. His hypothesis is based on the work of Princeton's John Wheeler, who had earlier predicted that turbulent fluctuations existed in the immediate aftermath of the Big Bang. Hawking reasoned that regions of heightened pressure caused by these pockets of turbulence could each have squeezed as much as a billion tons of matter into a space smaller than a subatomic particle, creating enormous numbers of what Hawking dubbed primordial black holes. And since these tiny objects would have been constantly losing mass since the birth of the universe—a process that would cause them to grow smaller and hotter all the time and thereby accelerate their rate of emission—they should by now have reached a point where they are radiating furiously enough to explode *(pages 111-115)*. No such explosions have been detected, which could mean that fewer primordial black holes were created than Hawking calculated, or that they will take longer to explode—or, perhaps, that they never formed at all.

TANTALIZING HINTS
Primordial black holes remain highly speculative, but astronomers have obtained promising evidence of the existence of their larger counterparts. Although a black hole cannot be observed directly, physicists began to speculate in the 1960s that the presence of a black hole could be inferred from its interactions with a nearby companion. In 1971, just such an arrangement was apparently detected in the constellation Cygnus. An invisible x-ray source, labeled Cyg X-1, was found near a bright blue supergiant. From Doppler-shift measurements of the supergiant, astrophysicists were able to determine not only that it was orbiting a common center of mass with an unseen companion but also that the companion's mass, though less than that of the supergiant, was at least eight times that of the Sun. Furthermore, the diameter of the x-ray source—estimated from a telltale flickering in the signal—seemed to indicate that this mass was far too dense for the companion to be a neutron star. In all likelihood, it was a black hole, and the x-rays were emanating from a stream of gas spiraling toward it from the outer layers of the supergiant and forming an accretion disk around it.

Even more enticing were observations made in 1982 from the Cerro Tololo Inter-American Observatory in Chile by astronomer Anne Cowley, then of the University of Michigan, and coworkers from the Dominion Astrophysical Observatory in British Columbia. This time, the x-ray source, located in the Large Magellanic Cloud, consisted of a visible star orbiting a more massive unseen companion. At ten times the mass of the Sun, LMC X-3, as the invisible partner was named, was far above a neutron star's theoretical limit of three solar masses. That such a massive object did not outshine its visible partner made LMC X-3 a very convincing black hole candidate.

Yet some astronomers continue to reserve judgment. If, for example, the x-rays from Cyg X-1 are emanating not from a surrounding disk of material

Black Holes from the Beginning of Time

At first, British astrophysicist Stephen Hawking did not want to believe his own calculations, which seemed to indicate that huge numbers of mini black holes might even now be exploding throughout the universe. The tiny black holes, measuring just trillionths of an inch across at their creation but containing millions of tons of mass, were Hawking's own theoretical brainchild, the product of his struggle to reconcile quantum mechanics (which describes the behavior of molecules, atoms, and subatomic particles) with Einstein's general theory of relativity (which describes the large-scale effects of gravitation). They would have come into being immediately after the Big Bang, when stupendous pressure crushed clumps of particles together.

As envisioned by Hawking, mini black holes (also called primordial black holes) possess the same gravitational properties as their stellar cousins, the collapsed remains of massive stars *(box, below)*—but on a far smaller scale. The size of both is defined by a boundary called the event horizon, marking off a sphere of gravitational influence so strong that it can swallow up light. The event horizon of a stellar black hole is tens of miles in diameter. In contrast, a primordial black hole with the mass of Earth, say, would have an event horizon only two-thirds of an inch across. An even smaller black hole, with the billion-ton mass of a small mountain, would be no bigger than a proton. Yet within its event horizon, this tiny cosmic fossil would be as mighty as any stellar black hole, clutching its contents in a grasp far stronger than any bond known in the familiar world.

Type of Black Hole	Mass	Diameter of Event Horizon	Distance at Which Gravity Equals Earth's
Mini	1 billion tons	.00000000000006 inch	9 feet
Intermediate	6 solar masses	11 miles	6 million miles
Supermassive	5 million solar masses	9,200,000 miles	5 billion miles

A primordial black hole is many orders of magnitude smaller in mass and volume than a black hole created by stellar collapse, but the two theoretical objects share the same anatomy. At the center is a compressed, infinitely dense mass known as a singularity, surrounded by an imaginary sphere, or event horizon. Anything falling inside the event horizon—within a distance from the singularity known as the Schwarzschild radius—is held there by gravity.

An Unexpected Source of Radiation

In 1974, contrary to accepted theory regarding the overpowering nature of black holes, Hawking found mathematical evidence that particles could, in fact, make their way out of the singularity's gravitational domain. The escape route was provided by a kind of loophole in quantum mechanics.

Quantum physicists view "empty" space as a medium filled with so-called virtual particles, curious objects that are not directly observable, or "real," but rather pop into and out of existence in as little as a billionth of a trillionth of a second. Virtual particles materialize seemingly out of nothingness in matter-antimatter pairs—a quark and an antiquark, for example, or a neutrino and an antineutrino—and then disappear again almost immediately in a process known as annihilation *(below, left)*. Because the particles appear for only the briefest flicker, they can be detected only indirectly, by their minute effects on other events. They have been found, for example, to produce small shifts in the spectrum of light from excited hydrogen atoms.

When a virtual pair comes into existence near the event horizon of a black hole, however, one member of the pair may be captured, swallowed up by the black hole's gravity *(below, right)*. Without a partner, the remaining particle cannot be annihilated—but it needs a source of mass to become "real." The particle thus steals a minute fraction of energy from the gravitational field of the black hole and becomes free to move away into space. The escape, in effect, constitutes an emission—something that runs counter to the original concept of black holes.

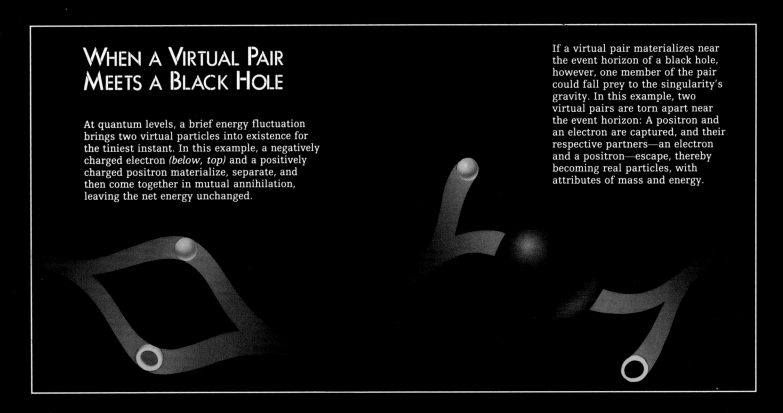

When a Virtual Pair Meets a Black Hole

At quantum levels, a brief energy fluctuation brings two virtual particles into existence for the tiniest instant. In this example, a negatively charged electron *(below, top)* and a positively charged positron materialize, separate, and then come together in mutual annihilation, leaving the net energy unchanged.

If a virtual pair materializes near the event horizon of a black hole, however, one member of the pair could fall prey to the singularity's gravity. In this example, two virtual pairs are torn apart near the event horizon: A positron and an electron are captured, and their respective partners—an electron and a positron—escape, thereby becoming real particles, with attributes of mass and energy.

Evaporating in a Blaze of Fireworks

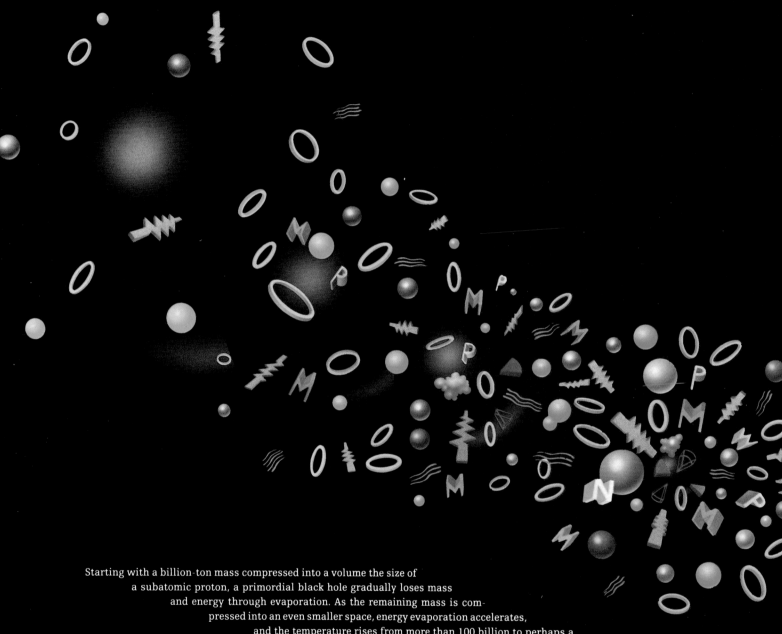

Starting with a billion-ton mass compressed into a volume the size of a subatomic proton, a primordial black hole gradually loses mass and energy through evaporation. As the remaining mass is compressed into an even smaller space, energy evaporation accelerates, and the temperature rises from more than 100 billion to perhaps a billion trillion trillion degrees Kelvin. Beyond a certain point of mass depletion, the remaining mass can no longer hold together gravitationally, and the exploding black hole becomes a kind of cosmic gusher, spewing matter and energy into the universe.

According to Hawking's theory, all black holes are depleted over time by the leakage of mass and energy, but for holes that are the remnants of massive stars, the loss is scarcely noticeable. In the case of a mini black hole, however, the depletion process, known as evaporation, whittles inexorably at the hole's relatively small amount of mass. As the hole shrinks, its temperature rises, increasing the hole's radiation of energy until, finally, the black hole's gravity releases its iron grip and all remaining mass spews out. A black hole having an original mass of, say, a billion tons would evaporate in a blaze of energy equivalent to the explosion of a million billion tons of dynamite.

Given the presumed age of the universe and the evaporation rate of black holes, such blasts should be happening even now—a fireworks display that would be visible from Earth as a flash of gamma radiation. Gamma ray bursts, easily detectable with an orbiting telescope, would provide the first observational evidence that mini black holes exist, but the odds of catching one in the act are long: Hawking estimates that within one light-year of the Solar System, mini black holes explode at the rate of only one or two per century, at most.

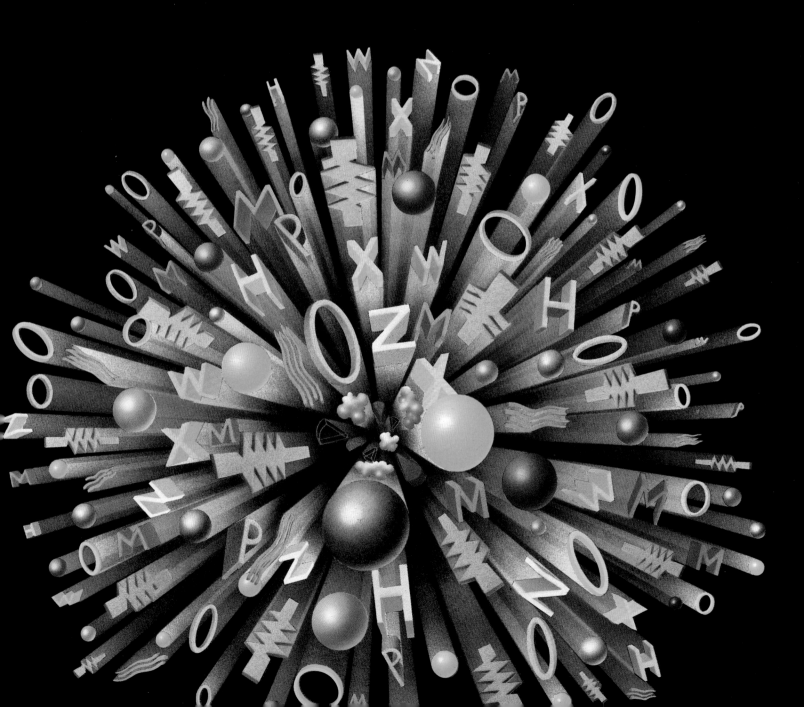

but from one small patch of superheated gas, the unseen companion could be much larger in diameter and therefore not dense enough to be a black hole. The same might hold true for LMC X-3. Although these interpretations are unlikely, they have prevented even the most fervent black hole advocates from claiming definitive proof. It seems that before the issue is truly settled, observers will have to uncover examples in which a black hole represents the only possible explanation.

ON A GALACTIC SCALE

Meanwhile, the discovery in the early 1960s of quasars—powerful sources of radiation shining from the universe's most distant reaches—had led theorists to propose still another kind of black hole. In 1967, Cambridge's Donald Lynden-Bell suggested that the energy pouring from quasars could result from the presence of gigantic black holes containing millions of solar masses. Such supermassive black holes might have formed all at once from the collapse of huge gas clouds. Or they might have developed in piecemeal fashion, amalgamated from stars or individual black holes of more conventional size. In either case, a process akin to what is happening in Cyg X-1 and LMC X-3 seems to be at work, though on a gargantuan scale. Rather than sucking gas from the atmosphere of a companion, the enormous gravity of one of these behemoths would attract whole stars, tearing them to gaseous shreds before devouring them. The gas would spiral around the black hole's event horizon, and the pressure and friction thus created would heat it up to millions or billions of degrees; glowing white-hot, it would radiate so brightly that it could be seen on the other side of the universe.

Shortcuts in space-time. Theoretical physicist John Wheeler *(left)*, who coined the term black hole in 1967, presents a fanciful discussion of some of cosmology's fundamental mysteries to a group of colleagues at Princeton University. In 1956, Wheeler developed equations describing so-called quantum fluctuations that might give rise to wormholes—hypothetical tunnels connecting discrete space-time regions. Caltech theorist Kip Thorne *(below)* later devised solutions for wormholes that humans might be able to traverse.

Although quasars continue to puzzle astronomers, black holes remain the most appealing explanation. For one thing, some quasars vary in brightness from time to time, and the period of the variations can be as short as a day. Whatever the reason for these changes in brightness, the speed at which they occur proves that the quasars must be relatively small: Since nothing can move faster than light, the surface over which the changes are spreading must be no more than a light-day in breadth (about 16 billion miles, or the size of the Solar System). Getting the observed energy out of so small a region suggests a concentration of matter that only a black hole could muster. Another clue is that quasars appear to be the bright centers of young galaxies, where there would be plenty of matter to feed a black hole. Many astronomers assume that quasarhood represents a period of galactic adolescence, and that after a galaxy's core material has been vacuumed up by the central black hole, the galaxy settles down into a kind of conservative middle age; the black hole remains, but there is less matter nearby for it to consume and thus no quasarlike glow. Recent evidence that even the most staid and stable galaxies, such as Andromeda and the Milky Way itself, may harbor black holes at their cores supports such a view.

THEORETICAL EXTREMES
Some theorists, however, have proposed an even more outlandish explanation for the violent energy emanating from quasars. These researchers suggest that quasars may be the exact physical opposites of black holes. The idea has its foundation in the mind-bending principles of relativity. Since time is just another dimension, there is no intrinsic rule that objects have to go forward in time, any more than there is a rule that objects always have to move in any particular spatial direction. The universe, of course, does appear to go forward in time, but this undeniable fact has so far eluded mathematical explanation.

In purely theoretical terms, then, the equations of relativity work just as well if time is reversed, which means that if black holes are possible, so are their opposites. Known as white holes, each of these peculiar beasts would begin as a singularity rather than end as one. And instead of gathering in large amounts of mass and energy, a white hole would, in effect, eternally gush. (Although they are similar to Hawking's exploding primordial black holes in some respects, white holes involve an entirely different process, unrelated to quantum mechanics.)

Could quasars be white holes? Although the idea achieved some popularity during the 1970s, the majority of astrophysicists are much more comfortable with black holes. For one thing, it is relatively easy to understand how black-hole quasars could have existed early on and then died out, while the same does not apply to white holes. One other phenomenon, however, seems to fit the description of a white hole in a general sense. The Big Bang began as a singularity that expanded inexorably outward, creating an entire universe in the process. But this concept has also been criticized as inappro-

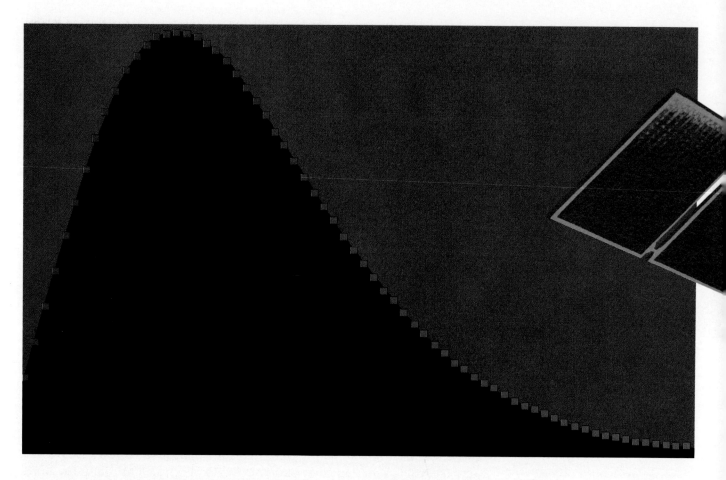

priate: The formulas describing white holes involve the spewing of matter and energy into a region of space, whereas during the explosive birth of the universe, it was space itself that was expanding.

Even if white holes are merely a mathematical construct, an obvious question about them is where their matter would come from. One potential answer emerges from a closer look at the equations of relativity. In the 1930s, Einstein and American theoretical physicist Nathan Rosen, who worked with Einstein at Princeton, discovered that singularities had an odd effect on the fabric of space-time. They realized that the gravity well in which a singularity resides would be bottomless, either opening onto some other hypothetical universe or connecting to another region of this universe's space-time. First known as an Einstein-Rosen bridge, the connection was later dubbed a wormhole by John Wheeler, again proving himself a master deviser of cosmological nomenclature. The term applies by analogy: Just as a worm's path through an apple can serve as a shortcut from one side of the apple to the other, so can a wormhole in space-time connect two widely separated regions. And it might provide a conduit through which the matter crushed to nothingness in a black hole could burst forth in another part of the universe.

Science-fiction writers and physicists alike have been intrigued by the possibilities. When he was conceiving his first novel, *Contact,* Cornell astronomer Carl Sagan decided that he would have his characters travel the

A plot of data from the Cosmic Background Explorer *(COBE),* a satellite lofted into space in late 1989 carrying three different sensing instruments, shows almost perfect agreement between measurements of microwave and infrared radiation *(small boxes)* and a smooth curve predicted for so-called blackbody radiation at a temperature of 2.73 degrees Kelvin. The finding supports the Big Bang theory for the origin of the universe some 15 billion years ago, since it reflects expected readings for the faint echoes reverberating from that cosmic explosion. However, the smooth background radiation provides no evidence for irregularities in the early universe that might have led to the creation of galaxies, leaving the observed distribution of galaxies unexplained.

cosmos by way of wormholes. He wanted to be particularly scrupulous about his physics, so he got in touch with Kip Thorne, a former student of John Wheeler's and a premier theoretical physicist in his own right. As Thorne was well aware, there was a fundamental problem: The same equations that describe the existence of wormholes predict that a wormhole opening would be far smaller than a subatomic particle. Furthermore, even if it were somehow possible to create a larger wormhole, all the mathematics indicated that it would pinch shut an instant after it formed, too fast for a spaceship to pass through.

Undeterred, Thorne enlisted the help of Michael Morris, one of his graduate students at Caltech, to find out how a wormhole might be propped open. They constructed a solution *(pages 128-129)* that admittedly relied on several unknowns but nonetheless proved that it was theoretically feasible to traverse an artificially created wormhole, a possibility that Einstein's original conception of a wormhole had disallowed. Inspired by their success, they took the argument a step further. Imagine, they said, that one end of the wormhole was somehow set moving at 99 percent of light speed for a period of, say, twenty years and then brought back to its starting point. Since, according to Einstein, time slows down for fast-moving objects, the time frame within that mouth of the wormhole would be more than 19 years behind the time frame just outside it. Therefore, a spaceship entering that mouth would emerge from the other mouth long before it began the trip. Thorne left the task of engineering this time machine to an "arbitrarily advanced civilization," his way of saying that although the laws of physics apparently permit such a phenomenon, no one can say whether it could in fact be accomplished. His point was to push Einstein's equations to their limit and see what resulted.

Other theorists have taken the concept of wormholes even further. Wheeler again has played a significant role, by originating the idea that the fabric of space-time itself is subject to the same kind of quantum fluctuations that affect the behavior of electrons and make them so hard to pin down. Building on this notion, Hawking has proposed that wormholes are forming all the time, everywhere in the cosmos, as tiny bulges on the surface of space-time. And before it pinches off, each one gives rise to a brand-new baby universe, permanently inaccessible but capable of growing into a full-fledged cosmos. Indeed, speculates Hawking, the known universe may have come into being in the same way, a tiny spark bursting into existence from a previous cosmos. Any new cosmos would in turn create countless universes of its own, and each of these would do the same, generating an infinitely complex lacework of budding worlds.

AN ALTERNATE MODEL
Even while physicists such as Hawking pursue the esoteric possibilities of their theories, new questions arise about some of the fundamental tenets of cosmology. The Big Bang model itself, still firmly entrenched as the most plausible explanation of how the universe began, has not been immune to criticism. One theorist in particular has developed an alternative hypothesis that exhibits an appealing simplicity. Although the idea has been branded as preposterous by most astrophysicists, it has, at the very least, stirred scientists to reexamine some of their long-standing assumptions.

In November of 1989, NASA launched the Cosmic Background Explorer *(COBE)*, the first satellite specifically designed to measure the Big Bang's leftover radiation, a faint microwave glow that pervades all of space. Earth-based measurements had already shown that this radiation is surprisingly even. Equipped with three different types of extremely sensitive detectors, *COBE* would attempt to pick up slight fluctuations that most theorists feel are necessary to explain the observed clumping of galaxies into such structures as the Great Wall and the Great Attractor *(pages 81-83)*.

The initial results were both satisfying and perplexing. On the one hand, the readings seemed to confirm that the Big Bang actually happened. On the other hand, the radiation showed not the tiniest hint of any kind of unevenness, perfectly matching the theoretical ideal *(page 118)*. And without some sort of turbulence or density fluctuations in the early universe, it was hard to explain how gravity could have shaped galaxies into clusters, superclusters, and even larger agglomerations.

Swedish plasma physicist Hannes Alfvén, however, has an explanation. Since the 1930s, Alfvén has been working to explain galaxy clumping not through mathematical proofs but by laboratory experiment. His conclusion is that the primary shaping mechanism in the universe is not gravity but electromagnetism, an inherently much more powerful force. His laboratory

New recipe for the universe? At the heart of the Milky Way, filaments of plasma 130 light-years long (upper left in this radio image) lend support to arguments for an alternate theory for the origins of the universe. Plasma, a gas of electrically charged particles, makes up 99 percent of observed matter, and the dynamics of huge plasma structures—influenced as much by electromagnetic forces as by gravity—have led some physicists to question the gravity-based equations that underlie the Big Bang theory.

work has shown that plasma, or ionized gas, can be shaped into long, twisted filaments under the influence of electrical currents and their accompanying magnetic fields. And if these effects can be produced in the lab, he reasons, why not in space itself?

In the 1960s, satellites proved that interplanetary space is filled with such currents and fields and that tenuous filaments of plasma exist throughout the Solar System. Furthermore, radio observations in 1984 uncovered evidence of huge filaments emanating from the center of the Milky Way, structures strikingly similar to those in a computer simulation designed by Anthony Peratt, a former student of Alfvén's. Current estimates that more than 99 percent of the universe's visible matter is in the form of plasma seem to lend even further support.

Alfvén and those few scientists who agree with him believe that all the accumulations of matter in the universe, from stars and planets to galactic clusters stretching for millions of light-years, result from the effects of magnetic fields induced by electrical currents flowing through plasma. As for the universe's origin, Alfvén notes, "There is no rational reason to doubt that the universe has existed indefinitely, for an infinite time."

The skeptics remain unconvinced. Despite the evidence in its favor, the theory abounds with problems. For one thing, effects produced in the laboratory do not of necessity translate to the infinitely larger scale of the universe. On the contrary, although electromagnetism is intrinsically more powerful than gravity, over great distances electrical charges tend to cancel each other, while gravitational forces continue to exert an influence even across millions of light-years. Furthermore, the filamentary structures generated by Peratt's computer model that match those found at the heart of the Milky Way have been produced by other simulations based on gravity alone. Also, Alfvén's plasma cosmology has trouble explaining such observed realities as galaxy recession and the microwave background radiation.

Nevertheless, Alfvén and Peratt have held to their maverick views, noting that many aspects of the Big Bang scenario are just as speculative and open to question. As Alfvén himself notes, "The significant point is that there are alternatives to the Big Bang." As astronomers struggle to understand a still-baffling cosmos, a willingness to challenge accepted theories and expand the limits of imagination may prove indispensable.

LEAPS ACROSS SPACE-TIME

The equations of the general theory of relativity published by Albert Einstein in 1915 expressed a revolutionary view of reality, replacing the familiar three-dimensional framework of physical events with a four-dimensional geometry—three dimensions of space and one of time. To picture the workings of this Einsteinian universe, four-dimensional space-time can be envisioned as a two-dimensional rubber sheet *(right)*. If a steel ball representing a massive star were placed on the sheet, it would create a depression. A marble rolling on the sheet's curved contours would—as if drawn by a mysterious power—fall in toward the "star," orbiting it like a planet. According to Einstein, gravity is nothing more than the geometry of space-time itself.

This geometry also affects light, whose speed—186,000 miles per second—is one of the few constants in Einstein's universe. The effect may be understood by imagining two pulses of light approaching an observer from equidistant points in the universe. One would expect the pulses to arrive simultaneously. However, one pulse passes close to a star and so follows the curvature of space-time induced there by the stellar mass; the other travels through the relatively flat contours of empty space-time. Thus, as gauged by the observer, the star-grazing light pulse takes longer to arrive.

Physicists are still working out the implications of Einstein's general theory. Among the phenomena it predicts are black holes—regions in space-time resulting from the collapse of supermassive stars; the gravity of a black hole is so intense that not even light can escape. Stranger still are wormholes, shortcuts between far-distant reaches of space and time. If they exist, some may permit time travel—a prospect that raises fascinating questions about cosmic order, causality, and the laws of nature.

Tunnel Connectors

Using the field equations of general relativity, scientists have developed a mathematical narrative that describes the typical genesis of a black hole. The process begins when a massive, aging star, its supply of nuclear fuel spent, no longer produces sufficient heat and outward pressure to counteract the gravitational pull of its own particles. Catastrophic stellar collapse results: The once-bright giant draws inward, shrinking ever smaller as its gravitational field builds. Finally, crushed to tremendous density, the star is extinguished altogether. In its place is the space-time abyss known as a black hole.

At the black hole's center is a singularity—a vanishing point whose density is so enormous that it causes space-time to turn in upon itself, like a serpent swallowing its tail. Here, all the usual laws of physics cease to apply. Astrophysicists can only guess at the strange rules that prevail, but the possibilities are bizarre. Nothing would prevent the singularity from penetrating the very fabric of space and time.

Though impossible to depict in four dimensions, such an event can be approximated in two, as shown at right. Burrowing through the sheet of space-time, the singularity could forge a link—a wormhole—to another region of space-time. Like a tunnel through a mountain, this hypothetical connection would greatly reduce the distance between far-removed parts of the universe. In theory, photons, planets, and small-mass stars drawn in by the black hole's intense gravity might be siphoned through the wormhole and disgorged at the other end. Traveling via these cosmic thruways, matter might move faster than light journeying the long way around.

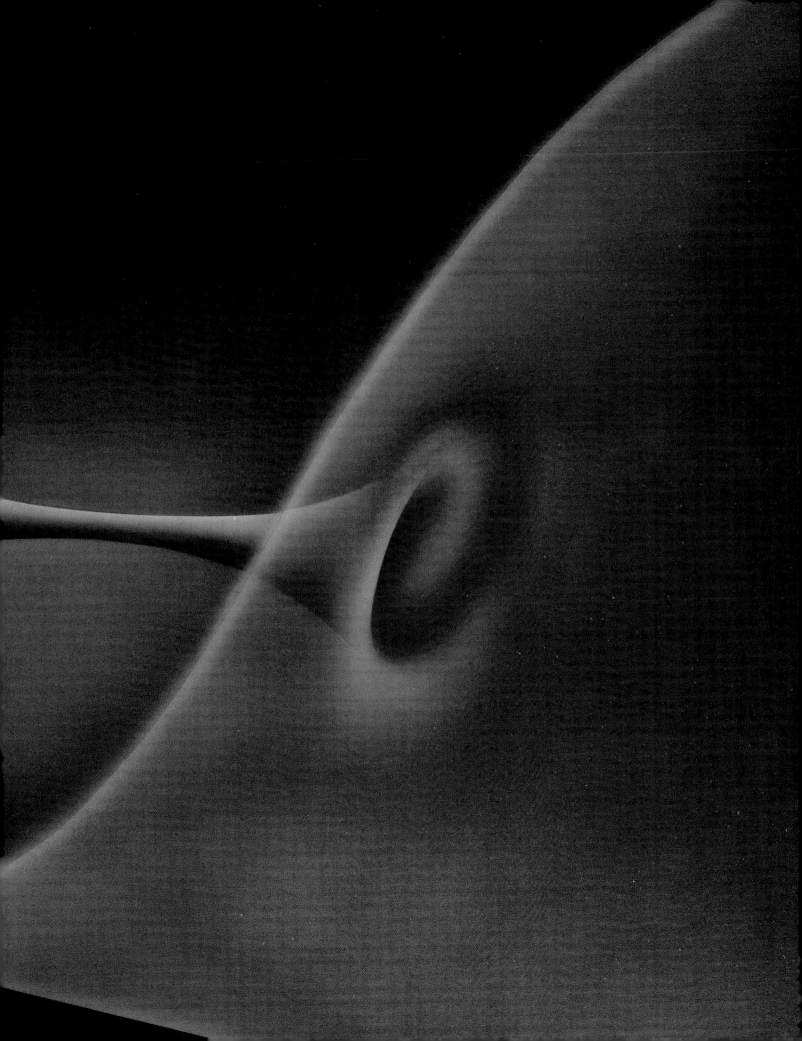

The Closing Door

Ever since the publication of the theory of general relativity, venturesome astrophysicists have engaged in a sort of mathematical hunt aimed at chasing down the various species of black holes and wormholes allowed by Einstein's equations. The first such investigative effort was the work done by the German astronomer Karl Schwarzschild in 1916 describing a wormhole that extends from a nonrotating, perfectly spherical black-hole mass. Later calculations revealed that any space traveler attempting to pass through a Schwarzschild wormhole would be crushed by a contraction in the wormhole's throat—a kind of gravitational hiccup *(right)*.

Such troubles are avoided in a black-hole scheme devised in 1918 by Heinrich Reissner of Germany and Gunnar Nordstrom of Finland. According to their theory, an electrical or magnetic charge lacing the black hole generates a repulsive force that stems internal gravitational tides, holding the wormhole open for safe passage into a labyrinth of past and future universes. But Reissner-Nordstrom wormholes are unstable and short-lived: Even a photon traversing such a wormhole would cause it to collapse, and in any case, no black hole could retain a charge for long in this charge-balanced universe.

In yet another scenario, worked out in 1963 by New Zealand mathematician Roy Kerr, the black hole's mass spins, giving rise to a doughnut-shaped singularity. Spaceship passengers traveling through the ring would risk being irradiated, but if they somehow got through, they would enter a region where they could travel back in time. Unfortunately, a Kerr black-hole wormhole is prone to the same instabilities as its Reissner-Nordstrom cousin.

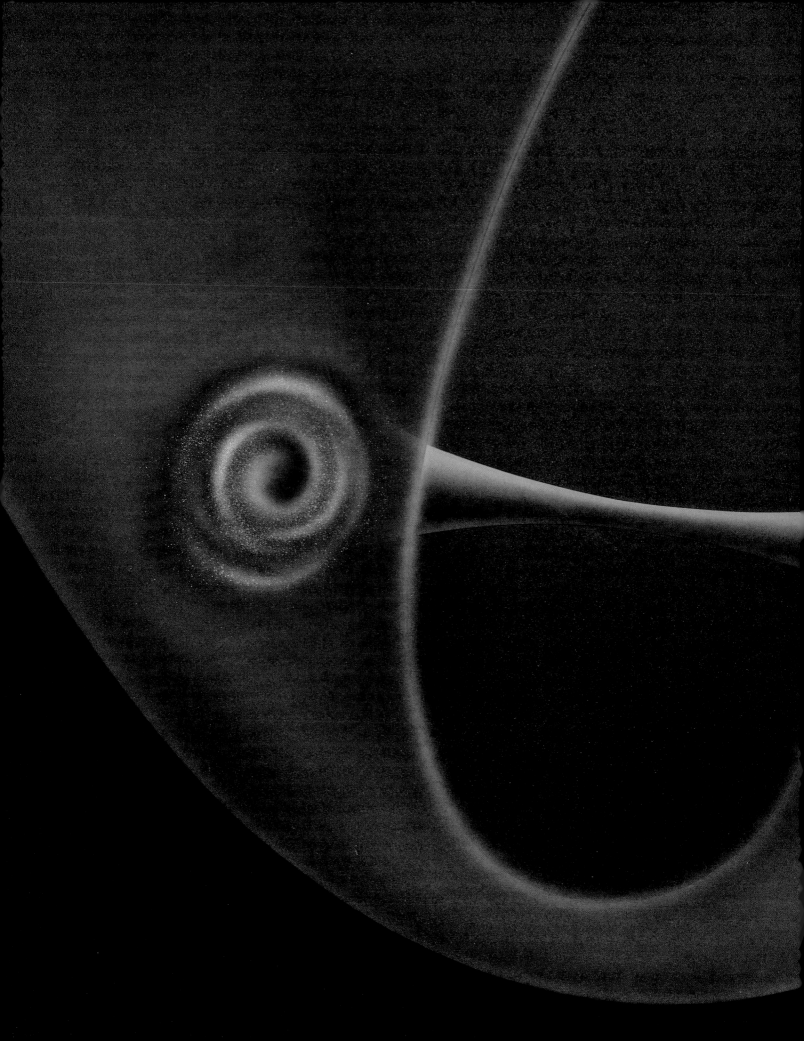

An Exotic Solution

Hypothetically, if a way could be found to keep wormholes from clamping shut like the jaws of some cosmic snapping turtle, then all of space and much of time would be within a traveler's reach. Recently, two American relativity theorists, Michael Morris and Kip Thorne, proposed what may be a possible—albeit wildly conjectural—solution: a balm of "exotic matter" that, when applied to the wormhole's throat, would quiet the gravitational spasms and hold the star gate open.

Morris and Thorne summoned exotic matter from an imaginary camp of particles allowed by quantum theory. If exotic matter exists, it would be ultradense and likely induce phenomenal stresses in anything that contacted it. Consequently, wormhole engineers intent on affixing it to the wormhole's throat and tunnel *(left)* would have to insulate themselves and travelers from exposure, perhaps by passing a vacuum tube down the wormhole's throat. Thus protected, wormhole traffic could safely voyage between distant points in the universe in a twinkling.

Conceivably, exotic-matter wormholes might also be converted into time machines. Using something massive like a neutron star as a sort of gravitational hook, engineers could drag one wormhole mouth to and fro at near light speed. According to Einstein's so-called special theory, the relativistic motion would cause the accelerated mouth to age more slowly than the stationary one. A spaceship entering the younger mouth would travel back in time as it passed through the wormhole—but only as far back as the moment when the mouth was first accelerated.

Paths to Everywhere

One possible place to find wormholes, scientists say, is in supertiny bits of space-time. On a scale trillions of times smaller than an atom, space-time loses its atomic-level smoothness: A host of high-energy quantum particles trigger huge energy fluctuations that, like mass, warp space-time's geometry, transforming it into a kind of seething foam. Physicists speculate that some of the more violent fluctuations may puncture space-time, creating Lilliputian wormholes. According to mathematical physicist Stephen Hawking, quantum fluctuations may also spawn minuscule bubbles that, over time, balloon into star-strewn universes billions of light-years across. As suggested by the illustration at left, these otherwise separate realms could be linked by an infinitude of wormholes. Harvard's Sidney Coleman has proposed that such quantum connections might constitute an intercosmic communications network. Within it, particle messengers tunneling between discrete—and perhaps very different—universes could influence each other's physical state, just as a photon's interaction with an electron can alter the electron's charge. The particle exchanges would take place with such rapidity and randomness that the normal orderliness of time and space would, in effect, be scrambled: Such a multicosmos dialogue would result in common physical values for the constants of nature. That is, the physical properties of any universe at any moment would be an average of all universes.

GLOSSARY

Absorption line: a dark line or band at a particular wavelength on a spectrum, formed when a substance between a radiating source and an observer absorbs electromagnetic radiation of that wavelength. Different substances produce characteristic patterns of absorption lines.

Accretion disk: a disk formed from gases and other materials drawn in by a compact body, such as a black hole or neutron star, at the disk's center.

Active galaxy: a galaxy with a highly energetic nucleus.

Angular momentum: a measure of an object's inertia, or state of motion, about an axis of rotation.

Antimatter: matter made up of antiparticles. Antiparticles are identical in mass to matter particles but opposite to them in properties such as electrical charge.

Atom: the smallest component of a chemical element that retains the properties associated with that element. Atoms are composed of protons, neutrons, and electrons; the number of protons determines the identity of the element.

Axion: a hypothetical, cold, dark particle proposed in certain theoretical models of the early universe. Axions are sometimes suggested as components of dark matter.

Big Bang: according to a widely accepted theory, the primeval moment, 15 to 20 billion years ago, when the universe began expanding from a state of infinite density.

Binary stars: a gravitationally bound pair of stars in orbit around their mutual center of mass. Binary stars are extremely common throughout the cosmos, as are systems of three or more stars.

Blackbody: an object, defined as a hypothetical ideal, that absorbs and may reemit all radiation reaching it. Radiation reemitted from a blackbody follows a characteristic thermal curve.

Black hole: theoretically, an extremely compact body with such great gravitational force that no radiation can escape from it. Proposed varieties include primordial, or mini, black holes, low-mass objects formed shortly after the beginning of the universe; stellar black holes, which form from the cores of very massive stars that have gone supernova; and supermassive black holes, equivalent to millions of stars in mass and located in the centers of galaxies.

BL Lacertae object: a galaxy with a particularly bright nucleus that changes rapidly in luminosity; named for Lacerta, the constellation in which the first such galaxy was identified. BL Lacertae objects are sometimes termed blazars.

Bridge: a filament of stars and gas stretching between two galaxies that may result from the galaxies' interaction.

Brown dwarf: a dim body of less than 0.1 solar mass without enough self-gravity to fuse hydrogen to helium.

Bubble chamber: a research device that maintains a quantity of liquid hydrogen or helium just at the point of boiling, so that high-energy particles passing through it will leave a trail of bubbles.

cD galaxy: a very bright, very large elliptical galaxy.

Cepheid variable: a star that changes regularly in luminosity over a set period of days or weeks.

Chaos theory: the study of randomness generated by fixed mathematical principles. In chaotic systems, a minor change in initial conditions produces exponential effects in outcome.

Charge-coupled device (CCD): an electronic array of detectors, usually positioned at a telescope's focus, for registering electromagnetic radiation.

Charged particle: an elementary particle having an electric or magnetic field that influences its interaction with other particles. For example, electrons carry a negative charge, protons a positive charge.

Cosmic background radiation: a detectable, steady emission of microwave radiation from all directions of the sky, commonly attributed to the Big Bang.

Cosmic ray: an atomic nucleus or charged particle moving at close to the speed of light; thought to originate in supernovae and other violent celestial phenomena.

Dark matter: a form of matter that has not yet been directly observed but whose existence is deduced from its gravitational effects.

Electromagnetic radiation: waves of electrical and magnetic energy that travel through space at the speed of light.

Electromagnetism: the force that attracts oppositely charged particles and repels similarly charged particles. Electromagnetism does not affect neutral particles such as neutrinos.

Electron: a negatively charged particle that normally orbits an atom's nucleus but may exist in isolation.

Elliptical galaxy: a galaxy shaped like a flattened sphere, with no discernible features and no disk.

Emission line: a bright band at a particular wavelength on a spectrum, emitted directly by the source and indicating by its wavelength a chemical constituent of that source.

Entropy: the measure of energy unavailable for doing work in a system or in the universe; increasing entropy implies a progression toward disintegration and randomness in the universe.

Ergosphere: the region around a black hole, outside the event horizon, where only objects that remain in motion can avoid entering the singularity.

Event horizon: the spherical boundary around a black hole's singularity, within which gravitational forces prevent anything, including light, from escaping.

Exotic matter: theoretical particles, including gravitons, axions, and magnetic monopoles, invoked to explain certain observed effects of matter.

False vacuum: in the theory of inflationary cosmology, an unstable state existing for a brief time immediately after the Big Bang when matter was able to expand at tremendous speed.

Frequency: the number of oscillations per second of an electromagnetic (or other) wave. *See* Wavelength.

Fusion: the combining of two atomic nuclei to form a heavier nucleus, releasing energy as a by-product.

Galaxy: a system of stars, gas, and dust that contains millions to hundreds of billions of stars.

Gamma ray: the most energetic form of electromagnetic radiation, with the highest frequency and the shortest wavelength.

General relativity: a theoretical account of the effects of acceleration and gravity on the motion of bodies and the observed structure of space and time.

Gravitational field: a region of space through which the gravity of a mass exerts measurable influence.

Gravitational lens: an optical effect of general relativity in which the gravity of a very massive body bends the light of an object behind it, distorting its apparent image and often producing one or more duplicate images.

Gravity: the force responsible for the mutual attraction of separate masses.

Gravity wave: a theoretical perturbation in an object's

gravitational field that would travel at the speed of light. General relativity predicts that gravity waves may result from accelerating, oscillating, or violently disturbed masses, including black holes and cosmic strings.

Great Attractor: an enormous concentration of matter in the region of space beyond the Hydra-Centaurus supercluster, suggested as the reason that some clusters of galaxies, including the Milky Way's own Local Group, apparently stream in that direction at great speeds.

Heisenberg uncertainty principle: the understanding that uncertain values are inevitable at the subatomic level since measuring techniques disrupt the particles being measured. For example, a particle's precise position and its velocity can never be known at the same time.

Helium: the second-lightest chemical element and the second-most-abundant; produced in stars by the fusion of hydrogen.

Herbig-Haro object: a clump of molecular gas that may mark shock waves where a jet emanating from a young star has hit a dense area of the interstellar medium.

Hubble flow: the motion of astronomical bodies receding from each other at a rate proportional to the distance between them—a consequence of the overall expansion of the universe. The Hubble constant is an estimate of the rate of expansion; values for the Hubble constant vary from 50 to 100 kilometers per second per megaparsec.

Hydrogen: the most common element in the universe. Stellar energy comes primarily from the fusion of hydrogen nuclei.

Hydroxyl: a molecule composed of one atom each of hydrogen and oxygen.

Inflationary cosmology: a theory accounting for the present form of the universe by proposing a sudden expansion of space that occurred at 10^{-35} seconds after the Big Bang.

Infrared: a band of electromagnetic radiation that has a lower frequency and a longer wavelength than visible red light.

Ion: an atom that has gained or lost one or more electrons and has become electrically charged. In comparison, a neutral atom has an equal number of electrons and protons, giving the atom a zero net electrical charge.

Jet: a thin, high-speed plasma stream ejected by a star, binary system, or galaxy. Many kinds of radiation, including radio waves and visible light, may be emitted by a jet or by interstellar or intergalactic matter that is disturbed by the jet.

Kelvin: an absolute temperature scale that uses Celsius degrees but sets 0 at absolute zero, or about -273 degrees Celsius.

Light-year: an astronomical distance unit equal to the distance that light travels in a vacuum in one year—almost six trillion miles.

Lobe: a vast expanse of plasma, highly luminous at radio or other wavelengths and emitted from the core of a radio galaxy.

Luminosity: an object's total energy output, usually measured in ergs per second.

Magnitude: a designation of an object's brightness or luminosity relative to that of other objects.

Maser: a source of radiation that focuses incoming energy into intense but variable radio beams; hydroxyl molecules and water molecules have been found to act as masers.

Mass: a measure of the total amount of material in an object, determined either by its gravity or by its tendency to stay in motion if in motion, or at rest if at rest.

Microwave: a radio wave of very high frequency and short wavelength.

Molecule: the smallest particle of an element or compound that retains its properties. A molecule consists of one or more atoms bonded together.

Nebula: a cloud of interstellar gas or dust; in some cases a supernova remnant or a shell ejected by a star. The term nebula once included all soft-edged objects, such as galaxies and globular clusters; this usage is no longer correct.

Nemesis: the name given to a hypothetical companion star of the Sun, invoked to account for periodic large-scale extinctions on Earth. According to this theory, comets disturbed by Nemesis's gravitational pull collided with Earth, raising lethal, long-lasting dust clouds that disrupted the ecology.

Neutral K meson: an elementary particle that decays into an unusual variety of forms, including pions, antineutrinos, and electrons.

Neutrino: a chargeless particle with little or no mass that moves at close to the speed of light.

Neutron: an uncharged particle with a mass similar to that of a proton; normally found in an atom's nucleus.

Neutron star: a very dense body composed of tightly packed neutrons; one possible product of a supernova explosion. Neutron stars are observed as pulsars.

Nova: a star that exhibits a sudden, temporary increase in brightness.

Orbit: the path of an object revolving around another object.

Parsec: an astronomical distance unit equal to approximately 3.26 light-years. The dimensions of galaxies are often given in kiloparsecs, or thousands of parsecs; larger distances are given in megaparsecs, or millions of parsecs.

Particle: the smallest component of any class of matter; for example, the elementary particles within an atom (such as electrons, protons, and neutrons); the smallest constituents of a gas (atoms and molecules); or the smallest forms of solid matter in space (interstellar and interplanetary dust particles).

Peculiar galaxy: a galaxy that has a more defined shape than a typically amorphous irregular but is not clearly a spiral or elliptical. Many peculiar galaxies are thought to result from a galactic interaction.

Peculiar motion: motion of stars and galaxies that is caused by the gravitational effects of nearby celestial masses and is unrelated to motion in the general expansion of the universe.

Pion: an elementary particle associated with the strong nuclear force.

Planet X: a hypothetical tenth planet orbiting beyond Pluto, proposed to explain disturbances in the orbits of Uranus and Neptune and their satellites.

Plasma: a gas of ionized particles, in contrast to ordinary gases, which are electrically neutral. Plasmas are sensitive to electrical and magnetic fields and are considered to be a fourth state of matter, along with ordinary gases, liquids, and solids.

Population III stars: a hypothetical class of very large stars, proposed to exist before the universe developed galactic structure.

Positron: an antiparticle to the electron, carrying a positive

electric charge.
Protogalaxy: a roughly spherical hydrogen cloud from which a galaxy forms; about thirty times the size of a mature galaxy.
Protostar: a large gaseous sphere, held together by its own gravitational attraction, that shrinks and compresses to become a star.
Pulsar: a radiating source that emits extremely regular bursts of energy at intervals of several seconds or less. Pulsars are almost certainly neutron stars. Millisecond pulsars emit bursts of radiation hundreds of times per second.
Quantum mechanics: a mathematical description of the rules by which subatomic particles interact, decay, and form atomic or nuclear objects.
Quasar: shortened from quasi-stellar radio source; an extremely powerful, bright source of energy, located in a very small region at the center of a distant galaxy, that outshines the whole galaxy around it.
Radiation: energy in the form of electromagnetic waves or subatomic particles.
Radio: the least energetic form of electromagnetic radiation, with the lowest frequency and the longest wavelength.
Radio galaxy: a galaxy that is an unusually strong source of radio waves and is distinguished from a radio-emitting quasar in that the source is not concentrated in a tiny region at the center.
Radio telescope: an instrument for studying astronomical objects at radio wavelengths.
Red giant: an aging, low-mass star that has greatly expanded and cooled down after consuming most of its core hydrogen.
Redshift: a stretching of the wavelengths of light, which shifts their spectral lines toward the red end of the spectrum. A Doppler redshift is caused by the motion of the light source; a cosmological redshift, by the expansion of space between the observer and the light source; and a gravitational redshift, by the time-distorting effects of the gravity of massive bodies.
Rotation: the turning of a celestial body about its axis.
Schwarzschild radius: the radius of a black hole's event horizon; anything falling within the Schwarzschild radius of a black hole cannot escape the hole's gravity.
Seyfert galaxy: an active disk galaxy with a very bright, starlike nucleus; first identified by American astronomer Carl Seyfert in 1943.
Singularity: the infinitely condensed mass at the center of a black hole. A singularity has no dimensions.
Solar flare: an explosive release of charged particles and electromagnetic radiation from a small area on the surface of the Sun.
Solar mass: a stellar mass unit equal to the Sun's mass, about two thousand trillion trillion grams.
Space-time: a four-dimensional concept of the universe that incorporates three spatial dimensions plus time.
Special relativity: a theory postulating that observers in uniform motion cannot perceive their motion and that all observers in such motion obtain the same value for the speed of light. From these two principles the theory concludes that measures of distance, time, and mass will vary depending on the motion of an observer moving uniformly in relation to the thing being measured.
Spectrum: the array of electromagnetic radiation, arranged in order of wavelength, from long-wave radio to short-wave gamma rays. Also, a narrower band of wavelengths, as the visible spectrum, in which light dispersed by a prism or other means shows its component colors; often banded with emission or absorption lines.
Starburst galaxy: a galaxy exhibiting the phenomenon of the sudden birth of many stars close together.
Strong nuclear force: the force that binds protons and neutrons together to form atomic nuclei.
Supercluster: a large association of clusters of galaxies.
Superluminal motion: motion at a speed that appears to exceed the speed of light. Theoretically impossible, such motion is usually attributed to illusions caused by the geometry of space.
Supernova: a stellar explosion that expels all or most of the star's mass and is extremely luminous.
Synchrotron radiation: a type of nonthermal radiation generated by electrons and other charged particles spiraling around magnetic field lines at near light speed.
T Tauri star: a very young star characterized by extensive and violent ejections of its mass.
Ultraviolet: a band of electromagnetic radiation with a higher frequency and a shorter wavelength than visible blue light. Because most ultraviolet is absorbed by the Earth's atmosphere, ultraviolet astronomy is normally performed in space.
Variable star: a star that changes in luminosity over time. Some variable stars change predictably and repeatedly; others change unpredictably, or only once.
Virtual particles: extremely short-lived particles created out of nothingness, as permitted by the uncertainty principle. Although they exist too briefly to be directly observed, the effects of their existence may be detected.
Wave: the propagation of a pattern of disturbance through a medium or space.
Wavelength: the distance from crest to crest or trough to trough of an electromagnetic or other wave. Wavelengths are related to frequency: The longer the wavelength, the lower the frequency.
Weak nuclear force: a very short range force responsible for particle decay.
White dwarf: an old, extremely dense star whose core has collapsed after exhausting its nuclear fuel of hydrogen and helium. A white dwarf with the mass of the Sun would be about the size of the Earth.
Wormhole: a hypothetical distortion in space-time linking widely separated black holes. Proposed varieties include Kerr wormholes, short-lived passages through hyperspace; Schwarzschild wormholes, which expand and contract quickly; and Reissner-Nordstrom wormholes, which are held open by the repulsive force of electric or magnetic charges.
X-ray: a band of electromagnetic radiation intermediate in wavelength between ultraviolet radiation and gamma rays. Because x-rays are completely absorbed by the atmosphere, x-ray astronomy must be performed in space.

BIBLIOGRAPHY

Books

Appenzeller, I., and C. Jordan (eds.). *Circumstellar Matter*. Dordrecht, Netherlands: D. Reidel, 1987.

Arp, Halton. *Quasars, Redshifts, and Controversies*. Berkeley, Calif.: Interstellar Media, 1987.

Bartusiak, Marcia. *Thursday's Universe*. New York: Times Books, 1986.

Baugher, Joseph F. *The Space-Age Solar System*. New York: John Wiley & Sons, 1988.

Boslough, John. *Stephen Hawking's Universe*. New York: William Morrow, 1985.

Brecher, Kenneth, and Michael Feirtag. *Astronomy of the Ancients*. Cambridge, Mass.: MIT Press, 1979.

Briggs, John, and F. David Peat. *Turbulent Mirror: An Illustrated Guide to Chaos Theory and the Science of Wholeness*. New York: Harper & Row, 1989.

Calder, Nigel:
Einstein's Universe. New York: Penguin Books, 1979.
The Key to the Universe. New York: Penguin Books, 1977.

Candee, Marjorie Dent (ed.). *Current Biography: Who's News and Why*. New York: H. W. Wilson, 1953.

Clark, David H. *The Quest for SS433*. New York: Penguin Books, 1985.

Close, Frank, Michael Marten, and Christine Sutton. *The Particle Explosion*. New York: Oxford University Press, 1987.

Cohen, Nathan. *Gravity's Lens: Views of the New Cosmology*. New York: John Wiley & Sons, 1988.

Contopoulos, G., and D. Kotsakis. *Cosmology: The Structure and Evolution of the Universe*. Trans. by M. Petrou and P. L. Palmer. Berlin: Springer-Verlag, 1984.

Cornell, James. *The First Stargazers*. New York: Charles Scribner's Sons, 1981.

The Cosmos (Voyage Through the Universe series). Alexandria, Va.: Time-Life Books, 1988.

Davies, Paul (ed.). *The New Physics*. Cambridge, England: Cambridge University Press, 1989.

The Far Planets (Voyage Through the Universe series). Alexandria, Va.: Time-Life Books, 1988.

Ferris, Timothy:
Galaxies. New York: Stewart, Tabori & Chang, 1982.
The Red Limit. New York: Quill, 1983.

Forward, Robert L. *Future Magic*. New York: Avon Books, 1988.

Gardner, Martin. *Time Travel and Other Mathematical Bewilderments*. New York: W. H. Freeman, 1988.

Giancoli, Douglas C. *Physics: Principles with Applications*. Englewood Cliffs, N.J.: Prentice-Hall, 1985.

Gillispie, Charles Coulston (ed.). *Dictionary of Scientific Biography*. New York: Charles Scribner's Sons, 1981.

Gleick, James. *Chaos: Making a New Science*. New York: Viking, 1987.

Gribbin, John:
Spacewarps. New York: Dell, 1983.
White Holes: Cosmic Gushers in the Universe. New York: Dell, 1977.

Hartmann, William K. *Astronomy: The Cosmic Journey*. Belmont, Calif.: Wadsworth, 1987.

Hawking, Stephen W.:
A Brief History of Time. New York: Bantam Books, 1988.
"The Edge of Spacetime." In *The New Physics*, ed. by Paul Davies. Cambridge, England: Cambridge University Press, 1989.

Hawking, Stephen W., and W. Israel (eds.). *Three Hundred Years of Gravitation*. Cambridge, England: Cambridge University Press, 1987.

Henbest, Nigel:
Mysteries of the Universe. New York: Van Nostrand Reinhold, 1981.
The New Astronomy. Cambridge, England: Cambridge University Press, 1983.

Herbert, Nick. *Faster Than Light: Superluminal Loopholes in Physics*. Markham, Ontario: Penguin Books Canada Ltd., 1988.

Hodge, Paul W. *Galaxies*. Cambridge, Mass.: Harvard University Press, 1986.

Hoffmeister, C., G. Richter, and W. Wenzel. *Variable Stars*. Trans. by S. Dunlop. Berlin: Springer-Verlag, 1985.

Hoyle, Fred, and Jayant Narlikar. *The Physics-Astronomy Frontier*. San Francisco: W. H. Freeman, 1980.

Kaler, James B. *Stars and Their Spectra: An Introduction to the Spectral Sequence*. Cambridge, England: Cambridge University Press, 1989.

Kaufmann, William J., III:
Black Holes and Warped Spacetime. San Francisco: W. H. Freeman, 1979.
Discovering the Universe. New York: W. H. Freeman, 1987.
Galaxies and Quasars. San Francisco: W. H. Freeman, 1979.
Relativity and Cosmology (2d ed.). New York: Harper & Row, 1977.
Universe. New York: W. H. Freeman, 1985.

Kaufmann, William J., III, et al. "The Black Hole." In *The Universe*, ed. by Byron Preiss and Andrew Fraknoi. New York: Bantam Books, 1987.

Kippenhahn, Rudolf. *Light from the Depths of Time*. Trans. by Storm Dunlop. Berlin: Springer-Verlag, 1984.

Lampton, Christopher. *Black Holes and Other Secrets of the Universe*. New York: Franklin Watts, 1980.

Lightman, Alan, and Roberta Brawer. *Origins: The Lives and Worlds of Modern Cosmologists*. Cambridge, Mass.: Harvard University Press, in press.

Littmann, Mark. *Planets Beyond: Discovering the Outer Solar System*. New York: John Wiley & Sons, 1988.

Maffei, Paolo. *Monsters in the Sky*. Trans. by Mirella and Riccardo Giacconi. Cambridge, Mass.: MIT Press, 1976.

Meyers, Robert A. (ed.). *Encyclopedia of Astronomy and Astrophysics*. San Diego, Calif.: Academic Press, 1989.

Mitton, Simon (ed.). *The Cambridge Encyclopaedia of Astronomy*. Cambridge, England: Cambridge University Press, 1985.

Mook, Delo E., and Thomas Vargish. *Inside Relativity*. Princeton, N.J.: Princeton University Press, 1987.

Moore, Patrick (ed.). *The International Encyclopedia of Astronomy*. New York: Orion Books, 1987.

Mundt, Reinhard. "Recent Observations of Herbig-Haro Objects, Optical Jets, and Their Sources." In *Circumstellar Matter*, ed. by I. Appenzeller and C. Jordan. Dordrecht, Netherlands: D. Reidel, 1987.

Nicolson, Iain. *Gravity, Black Holes and the Universe*. New York: John Wiley & Sons, 1981.

Pasachoff, Jay M., and Marc L. Kutner. *University Astronomy*. Philadelphia: W. B. Saunders, 1978.

Penrose, Robert. "Black Holes." In *The World of Physics*, ed. by Jefferson Hane Weaver. New York: Simon and Schuster, 1987.

Petit, Michel. *Variable Stars*. Trans. by W. J. Duffin. New York: John Wiley & Sons, 1987.

Raup, David M. *The Nemesis Affair: A Story of the Death of Dinosaurs and the Ways of Science*. New York: W. W. Norton, 1986.

Rees, Martin, Remo Ruffini, and John Archibald Wheeler. *Black Holes, Gravitational Waves and Cosmology: An Introduction to Current Research*. New York: Gordon and Breach Science, 1974.

Ronan, Colin A. *Deep Space*. New York: Macmillan, 1982.

Shipman, Harry L. *Black Holes, Quasars, and the Universe* (2d ed.). Boston: Houghton Mifflin, 1980.

Silk, Joseph. *The Big Bang*. New York: W. H. Freeman, 1989.

Snow, Theodore P. *Essentials of the Dynamic Universe*. St. Paul, Minn.: West, 1987.

Stars (Voyage Through the Universe series). Alexandria, Va.: Time-Life Books: 1988.

Stewart, Ian. *Does God Play Dice? The Mathematics of Chaos*. Cambridge, Mass.: Basil Blackwell, 1989.

Sullivan, Walter. *Black Holes: The Edge of Space, the End of Time*. Garden City, N.Y.: Anchor Press, 1979.

Thorne, Kip S., et al. "Black Holes: The Membrane Viewpoint." In *Highlights of Modern Astrophysics: Concepts and Controversies*, ed. by Stuart L. Shapiro and Saul A. Teukolsky. New York: John Wiley & Sons, 1986.

Toomre, Alar, and Juri Toomre. "Violent Tides between Galaxies." In *Readings from Scientific American: The Universe of Galaxies*, comp. by Paul W. Hodge. New York: W. H. Freeman, 1984.

Tucker, Wallace, and Karen Tucker. *The Dark Matter: Contemporary Science's Quest for the Mass Hidden in Our Universe*. New York: William Morrow, 1988.

Vehrenberg, Hans. *Atlas of Deep-Sky Splendors*. Cambridge, Mass.: Sky, 1983.

Vorontsov-Vel'yaminov, B. A. *Extragalactic Astronomy*. Trans. by Richard B. Rodman. London: Harwood Academic, 1987.

Wheeler, John A., and Remo Ruffini. "Introducing the Black Hole." In *The World of Physics*, ed. by Jefferson Hane Weaver. New York: Simon and Schuster, 1987.

Zeilik, Michael. *Astronomy, the Evolving Universe* (5th ed.). New York: John Wiley & Sons, 1988.

Zeilik, Michael, and Elske v. P. Smith. *Introductory Astronomy and Astrophysics* (2d ed.). Philadelphia: Saunders College Publishing, 1987.

Periodicals

Allman, William F. "Playing Fast and Loose with Time." *U.S. News & World Report*, December 19, 1988.

Andersen, Per H.:
"General Streaming of Galaxies Seen out to 100 Mpc." *Physics Today*, November 1986.
"Ripples Seen in the Universal Hubble Flow." *Physics Today*, October 1987.

Arp, Halton. "Ejection from the Spiral Galaxy NGC 1097." *Astrophysical Journal*, August 1, 1976.

"Astro-News: Violent Ejection Spews from Star." *Astronomy*, July 1982.

Bartusiak, Marcia:
"If You Like Black Holes, You'll Love Cosmic Strings." *Discover*, April 1988.
"Shifting Galaxies." *Omni*, May 1987.

Biretta, J. A., F. N. Owen, and T. J. Cornwell. "A Search for Motion and Flux Variations in the M87 Jet." *Astrophysical Journal*, July 1, 1989.

Blandford, Roger D., Mitchell C. Begelman, and Martin J. Rees. "Cosmic Jets." *Scientific American*, May 1982.

Blondin, John M., Arieh Konigl, and Bruce A. Fryxell. "Herbig-Haro Objects as the Heads of Radiative Jets." *Astrophysical Journal*, February 1, 1989.

Bothun, Gregory D., et al. "Discovery of a Huge Low-Surface-Brightness Galaxy: A Protodisk Galaxy at Low Redshift?" *Astronomical Journal*, July 1987.

Bridle, Alan H., and Richard A. Perley. "Extragalactic Radio Jets." *Annual Review of Astronomy and Astrophysics*, 1984.

Burbidge, Geoffrey. "Quasars, Redshifts, and Controversies." *Sky & Telescope*, January 1988.

Burns, Jack O. "Very Large Structures in the Universe." *Scientific American*, May 1986.

Burns, Jack O., and R. Marcus. "Centaurus A: The Nearest Active Galaxy." *Scientific American*, November 1983.

Burrows, Adam. "The Birth of Neutron Stars and Black Holes." *Physics Today*, September 1987.

Byrd, Deborah:
"Do Brown Dwarfs Really Exist?" *Astronomy*, April 1989.
"Starburst Galaxies." *Star Date*, September-October 1989.

"The Celestial Cartwheel." *Sky & Telescope*, December 1989.

Chandler, David L. "Alfven: Universe Is an Electric Place." *Boston Globe*, March 20, 1989.

Cline, Thomas L. "The Unique Cosmic Event of 1979 March 5." *Comments on Astrophysics*, 1980, Vol. 9, no. 1, pp. 13-22.

Cline, Thomas L., et al. "Detection of a Fast, Intense and Unusual Gamma-Ray Transient." *Astrophysical Journal*, April 1, 1980.

"COBE Mission Launched." *Astronomy*, February 1990.

Cowen, R. "Starlight Shadows Protogalaxy Finding." *Science News*, September 9, 1989.

Crane, Philippe, et al. "Cosmic Background Radiation Temperature at 2.64 Millimeters." *Astrophysical Journal*, November 1, 1989.

Croswell, Ken. "In Search of Planet X." *Odyssey*, May 1989.

Crutchfield, James P., J. Doyne Farmer, and Norman H. Packard. "Chaos." *Scientific American*, December 1986.

Darling, David. "Quest for Black Holes." *Astronomy*, July 1983.

Davis, Bob. "Two Astronomers at Cornell Claim They Found Galaxy in the Making." *Wall Street Journal*, August 29, 1989.

Davis, Marc, Piet Hut, and Richard A. Muller. "Extinction of Species by Periodic Comet Showers." *Nature*, April 19, 1984.

Davis, R. J., T. W. B. Muxlow, and R. G. Conway. "Radio Emission from the Jet and Lobe of 3C273." *Nature*, November 28, 1985.

DeWitt, Bryce S. "Quantum Gravity." *Scientific American,* December 1983.

Dickinson, Dale F. "Cosmic Masers." *Scientific American,* June 1978.

Dicus, Duane A., et al. "The Future of the Universe." *Scientific American,* March 1983.

Disney, Michael J., and Philippe Veron. "BL Lacertae Objects." *Scientific American,* August 1977.

"A Distorted View of the Southern Crab." *Science News,* January 13, 1990.

Dressler, Alan:
"In the Grip of the Great Attractor." *The Sciences,* September-October 1989.
"The Large-Scale Streaming of Galaxies." *Scientific American,* September 1987.

Ferris, Timothy. "Where Are We Going?" *Sky & Telescope,* May 1987.

Fisher, Arthur. "Red Spot Revealed." *Popular Science,* July 1988.

Flam, Faye. "Penned-In Positrons." *Science News,* May 11, 1989.

Freedman, David H. "Cosmic Time Travel." *Discover,* June 1989.

Fruchter, A. S., D. R. Stinebring, and J. H. Taylor. "A Millisecond Pulsar in an Eclipsing Binary." *Nature,* May 19, 1988.

Geballe, Thomas R. "The Central Parsec of the Galaxy." *Scientific American,* July 1979.

Geller, Margaret J., and John P. Huchra. "Mapping the Universe." *Science,* November 17, 1989.

Giovanelli, Riccardo, and Martha P. Haynes. "A Protogalaxy in the Local Supercluster." *Astrophysical Journal,* November 1, 1989.

Gleick, James. "New Images of Chaos That Are Stirring a Science Revolution." *Smithsonian,* December 1987.

Gorenstein, Paul, and Wallace Tucker. "Rich Clusters of Galaxies." *Scientific American,* November 1978.

Gregory, Stephen A.:
"Active Galaxies and Quasars: A Unified View." *Mercury,* July-August 1988.
"The Structure of the Visible Universe." *Astronomy,* April 1988.

Gregory, Stephen A., and Nancy D. Morrison:
"The Discovery of the Highest Redshift Galaxy We Know—The Largest Supercluster Filament." *Mercury,* March-April 1986.
"The Puzzle of Epsilon Aurigae: Results from the Recent Eclipse." *Mercury,* November-December 1986.
"Visible Synchrotron Emission from the Lobes of a Radio Galaxy." *Mercury,* July-August 1986.

Gribbin, John:
"Black Holes, White Holes and Wormholes." *Astronomy,* November 1976.
"Concept Opening Way to a Scientific World of Fantasy." *Smithsonian,* November 1977.

Harrington, R. S. "The Location of Planet X." *Astronomical Journal,* October 1988.

Hatzes, Artie P., G. Donald Penrod, and Steven S. Vogt. "Doppler Imaging of Abundance Features on Ap Stars: The Si II Distribution on γ^2 Arietis." *Astrophysical Journal,* June 1, 1989.

Hawking, Stephen W.:
"Black Hole Explosions?" *Nature,* March 1, 1971.
"The Quantum Mechanics of Black Holes." *Scientific American,* January 1977.
"The Quantum Wave Function of the Universe." *Science,* December 2, 1980.

Hilts, Philip J. "Birth of Galaxy Seen in Gas Cloud." *New York Times,* August 29, 1989.

Horgan, John:
"Cosmic News." *Scientific American,* March 1989.
"Shear Magic: Jupiter's Great Red Spot Is Conjured Up in Laboratories." *Scientific American,* May 1988.

"Imaging Distant Stars." *Sky & Telescope,* November 1986.

Ingersoll, Andrew P. "Fluid Mechanics: Models of Jovian Vortices." *Nature,* February 25, 1988.

Islam, Jamal N. "The Ultimate Fate of the Universe." *Sky & Telescope,* January 1979.

Kafatos, Minas, and Andrew G. Michalitsianos. "Symbiotic Stars." *Scientific American,* July 1984.

Kaiser, N. "Where Is the Great Attractor?" *Nature,* April 13, 1989.

Keel, William C. "Crashing Galaxies, Cosmic Fireworks." *Sky & Telescope,* January 1989.

Kerr, Richard A.:
"Does Chaos Permeate the Solar System?" *Science,* April 14, 1989.
"Pluto's Orbital Motion Looks Chaotic." *Science,* May 13, 1988.

Killian, Anita M. "Playing Dice with the Solar System." *Sky & Telescope,* August 1989.

Krauss, Lawrence M. "Dark Matter in the Universe." *Scientific American,* December 1986.

Kunzig, Robert. "The Wolf Effect." *Discover,* August 1988.

Lada, Charles J.:
"Cold Outflows, Energetic Winds, and Enigmatic Jets around Young Stellar Objects." *Annual Review of Astronomy and Astrophysics,* 1985.
"Energetic Outflows from Young Stars." *Scientific American,* July 1982.

Lemonick, Michael D.:
"Great Bubbles in the Cosmos." *Time,* November 27, 1989.
"Wormholes in the Heavens." *Time,* January 16, 1989.

Lerner, Eric J. "The Big Bang Never Happened." *Discover,* June 1988.

Lindley, David. "COBE Starts Its Search for Galactic Fingerprints." *Nature,* November 23, 1989.

Littmann, Mark. "Where Is Planet X?" *Sky & Telescope,* December 1989.

LoPresto, James Charles. "The Geometry of Space and Time." *Astronomy,* October 1987.

Lorre, Jean J. "Enhancement of the Jets in NGC 1097." *Astrophysical Journal,* June 15, 1978.

McClintock, Jeffrey. "Do Black Holes Exist?" *Sky & Telescope,* January 1988.

McKenzie, A.:
"Cloud Links Quasars to Seyfert Galaxies." *Science News,* September 30, 1989.
"Cosmic Cartographers Find 'Great Wall.'" *Science News,* September 25, 1989.

MacRobert, Alan. "Epsilon Aurigae: Puzzle Solved?" *Sky & Telescope,* January 1988.

Malin, David. "The Splendor of Eta Carinae." *Sky & Telescope,* January 1987.

Mallove, Eugene F. "The Self-Reproducing Universe." *Sky & Telescope,* September 1988.

Maran, Stephen P.:
"A Deviant Star." *Natural History,* June 1983.
"Slow-Motion Eclipse." *Natural History,* October 1983.
"Star Burst." *Natural History,* July 1983.
"Stellar Old-Timers." *Natural History,* February 1987.

Marcus, Philip S. "Numerical Simulation of Jupiter's Great Red Spot." *Nature,* February 25, 1988.

Meyers, S. D., J. Sommeria, and Harry L. Swinney. "Laboratory Study of the Dynamics of Jovian-Type Vortices." *Physica D,* 1989, Vol. 37, pp. 515-530.

Morris, Michael S., and Kip S. Thorne. "Wormholes in Spacetime and Their Use for Interstellar Travel: A Tool for Teaching General Relativity." *American Journal of Physics,* May 1988.

Muller, Richard A. "The Cosmic Background Radiation and the New Aether Drift." *Scientific American,* May 1978.

Nadis, Steve. "Galactic Tug-of-War." *Omni,* August 1988.

"News Notes: Puppis A: A Double Supernova?" *Sky & Telescope,* August 1989.

"News Notes: A Supernova Remnant's Jet." *Sky & Telescope,* February 1986.

Osmer, Patrick S. "Quasars as Probes of the Distant and Early Universe." *Scientific American,* February 1982.

Overbye, Dennis. "Out from under the Cosmic Censor: Stephen Hawking's Black Holes." *Sky & Telescope,* August 1977.

Palca, Joseph. "A Surprise Near Virgo." *Science,* September 1, 1989.

Parker, Barry:
"The First Second of Time." *Astronomy,* August 1979.
"Miniblack Holes." *Astronomy,* February 1977.

Penrose, Roger. "Black Holes." *Scientific American,* May 1972.

Peratt, Anthony L.:
"Guest Editorial: Electrical Engineering, Plasma Science, and the Plasma Universe." *IEEE Transactions on Plasma Science,* December 1986.
"Not with a Bang." *The Sciences,* January-February 1990.
"Plasma Cosmology, Part I: Interpretation of the Visible Universe." *The World & I,* August 1989.
"Plasma Cosmology, Part II: The Universe Is a Sea of Electrically Charged Particles." *The World and I,* September 1989.
"Space Plasmas." *The World & I,* March 1988.

Peterson, I.:
"Cosmic Evidence of a Smooth Beginning." *Science News,* January 20, 1990.
"Gigantic Gas Jet Points to Newborn Star." *Science News,* July 22, 1989.
"Jupiter's Spot of Order in Chaos." *Science News,* June 2, 1984.
"Looking Well Beyond the Great Attractor." *Science News,* April 15, 1989.
"A Model Spot for Jupiter." *Science News,* February 27, 1988.

Phinney, E. S. "Cosmic Merger Mania." *Nature,* August 24, 1989.

Pool, Robert:
"Chaos Theory: How Big an Advance?" *Science,* July 1989.
"Is It Chaos, or Is It Just Noise?" *Science,* January 6, 1989.

Poterma, Thomas A. "Guest Editorial: The Golden Anniversary of 'Magnetic Storms and the Aurorae.'" *IEEE Transactions on Plasma Science,* April 1989.

Powell, Corey S. "Astronomy Potpourri: Footballs, Eyeballs, Attractors and Twins." *Scientific American,* March 1990.

Price, Huw. "A Point on the Arrow of Time." *Nature,* July 20, 1989.

Price, Richard H., and Kip S. Thorne. "The Membrane Paradigm for Black Holes." *Scientific American,* April 1988.

Rees, M. J. "New Interpretation of Extragalactic Radio Sources." *Nature,* January 29, 1971.

Rees, Martin. "Galactic Nuclei and Quasars: Supermassive Black Holes?" *New Scientist,* October 19, 1978.

Reipurth, Bo. "The HH111 Jet and Multiple Outflow Episodes from Young Stars." *Nature,* July 6, 1989.

Rickey, Tom. "Seeing (Infra) Red: Astronomer Reaps Benefits of Developing New Technology." *Research Frontiers,* Fall 1989.

Roger, R. S., et al. "A Radio Jet Associated with the Supernova Remnant G332.4 + 0.1 (Kes 32)." *Nature,* July 1985.

Rothman, Tony. "God Takes a Nap: A Computer Finds That Pluto's Orbit Is Chaotic." *Scientific American,* October 1988.

Sawyer, Kathy. "9 Planet-Like Objects Reportedly Found." *Washington Post,* June 15, 1989.

Schorn, Ronald A. "The Extragalactic Zoo, I." *Sky & Telescope,* January 1988.

Schwarz, Hugo E., Colin Aspin, and Julie H. Lutz. "He 2-104: A Symbiotic Proto-Planetary Nebula?" *Astrophysical Journal,* September 1, 1989.

Schwarzschild, Bertram M.:
"Companion Galaxies Match Quasar Redshifts: The Debate Goes On." *Physics Today,* December 1984.
"Redshift Surveys of Galaxies Find a Bubbly Universe." *Physics Today,* May 1986.
"Why Is the Cosmological Constant So Very Small?" *Physics Today,* March 1989.

Shields, Gregory, A. "Behind the Scenes: Are Black Holes Really There?" *Astronomy,* October 1978.

Sienko, Tanya. "Bubbles in the Universe." *Space,* July-August 1989.

Silk, Joseph. "The Formation of Galaxies." *Physics Today,* April 1987.

Smith, David H.:
"Cosmic Fire, Terrestrial Ice." *Sky & Telescope,* November 1989.
"Mysteries of Cosmic Jets." *Sky & Telescope,* March 1985.

Struve, Otto. "Galaxies and Their Interactions." *Sky & Telescope,* February 1957.

Sussman, Gerald Jay, and Jack Wisdom. "Numerical Evidence That the Motion of Pluto Is Chaotic." *Science,* July 22, 1988.

Taubes, G. "The Mathematics of Chaos." *Discover,* September 1984.

Thorne, Kip S. "Gravitational Collapse." *Scientific American,* November 1967.

Trimble, Virginia. "Part Two: The Search for Dark Matter." *Astronomy,* March 1988.
Turner, Edwin L. "Gravitational Lenses." *Scientific American,* July 1988.
Verschuur, Gerrit L. "Is the Milky Way an Interacting Galaxy?" *Astronomy,* January 1988.
Visser, Matt. "Traversable Wormholes: Some Simple Examples." *Physical Review D,* May 15, 1989.
Waldrop, M. Mitchell:
"Are We All in the Grip of a Great Attractor?" *Science,* September 11, 1987.
"Astronomers Go Up Against the Great Wall." *Science,* November 17, 1989.
"Causality, Structure, and Common Sense." *Science,* September 11, 1987.
"COBE Confronts Cosmic Conundrums." *Science,* January 26, 1990.
"Gamma Rays from Cygnus X-1." *Science,* January 8, 1988.
"Heart of Darkness." *Science,* February 19, 1988.
"Pushing Back the Redshift Limit." *Science,* February 12, 1988.
"The Quantum Wave Function of the Universe." *Science,* December 2, 1988.
Weedman, Daniel. "Quasars: A Progress Report." *Mercury,* January-February 1988.
Whitmire, Daniel P., and Albert A. Jackson IV. "Are Periodic Mass Extinctions Driven by a Distant Solar Companion?" *Nature,* April 19, 1984.
Wilczek, Frank. "The Cosmic Asymmetry between Matter and Antimatter." *Scientific American,* December 1980.
Wilford, John Noble:
"Novel Theory Challenges the Big Bang." *New York Times,* February 28, 1989.
"Peering to Edge of Time, Scientists Are Astonished." *New York Times,* November 20, 1989.
Winkler, P. F., et al. "A Second Supernova Inside Puppis A?" *Nature,* January 5, 1989.
Wisdom, Jack. "Pluto's Orbital Motion Looks Chaotic." *Science,* May 20, 1988.
Wolstencroft, R. D., and W. J. Zealey. "The Peculiar Galaxy NGC 1097." *Monthly Notices of the Royal Astronomical Society,* December 1975.
"Wormholes and Time Machines." *Science News,* November 5, 1988.

Other Sources
Bridle, Alan. "Relativistic Jets and the Most Powerful Radio Sources in the Universe." Report no. 234. Brookhaven National Laboratory, May 20, 1987.
Cleggett-Haleim, Paula, and Carolynne White. "Early COBE Results in Accord with Big Bang Theory." News release. Washington, D.C.: NASA, January 13, 1990.
Forward, R. "Space Warps: A Review of One Form of Propulsionless Transport." Paper presented at the 25th Joint Propulsion Conference, July 10-12, 1989. Washington, D.C.: American Institute of Aeronautics and Astronautics.
Maran, Stephen P. Untitled news release. Washington, D.C.: American Astronomical Society, June 14, 1989.
Marcus, Philip S. "Nonlinear Mathematics and Jupiter's Great Red Spot." Unpublished manuscript. Department of Mechanical Engineering, University of California, Berkeley, no date.
Mather, John C. "Cosmic Background Explorer." Project report. Greenbelt, Md.: NASA Goddard Space Flight Center, no date.
NASA/Space Telescope Science Institute. "Nova Cygni 1975: Mystery Solved." News release. Baltimore, Md.: STSI, no date.
NASA/Space Telescope Science Institute. Photo release, no. PRC-9001C (Binary Quasar System, December 11-13, 1988). Baltimore, Md.: STSI, no date.
Pobojewski, Sally. "U-M Astronomer Says the Universe May Be Full of Dim 'Hidden' Galaxies Trapped Forever in an Early Stage of Galaxy Formation." News release. Ann Arbor: University of Michigan, June 12, 1989.
Visser, Matt. "Wormholes and Interstellar Travel." Unpublished manuscript. Los Alamos, N.Mex.: Los Alamos National Laboratory, March 1989.

INDEX

Numerals in italics indicate an illustration of the subject mentioned.

A
Abell, George: galaxy clusters identified by, 78, 79
Absorption-line patterns: of quasars, 70; of stars, 32
Active galaxies, 59, 60, 63-65
Alfvén, Hannes, 120-121
Alvarez, Luis, and son Walter, 24
Anderson, Carl David: work of, 97-98, 100
Andromeda (galaxy), 57-58, *94-95*
Antimatter and matter, 97-98, *99,* 100-101; virtual particles, pairs of, and black holes, 109, *112-113*
Arp, Halton, 62, 67-68, *69,* 76

Arp 297C (galaxy), *52-53*
Asteroids, orbital disruptions of, 23; and extinctions on Earth, 24, 26
Atlas of Peculiar Galaxies (Arp), 62

B
Baade, Walter, 107
Background radiation, cosmic: *COBE* measurement of, *118,* 120; hot spot in, 87, 89
Becklin, Eric, 29
Bessel, Friedrich Wilhelm, 31
Big Bang: alternative hypothesis to, 120-121; and matter-antimatter paradox, 100-101; and mini black holes, 110, 111; radiation left over from, 87, 118, 120; as white hole, 117-118

Binary quasar system, *109*
Binary star systems: Black Widow pulsar, 37, *48-49;* Epsilon Aurigae, 33, *35;* nova formation in, 51; R Aquarii, *30-31,* 32-33; Sirius A and B, *30,* 31-32; Southern Crab, theory of, *4-5*
Blackbody radiation: *COBE* measurement of, *118,* 120; hot spot in, 87, 89
Black holes, 101-117; anatomy of, 104, *111;* comparison of types, *chart* 111; electrically charged (Reissner-Nordstrom), 107; genesis of, 107-108, 124; Hawking's hypothesis, 109-110, 111, 112, 115; Laplace's concept, 102; Michell's concept, 101-102; mini (primor-

dial), 110, 111, *114,* 115; newspaper spoof, *105;* radiating, 109, *112-113;* reversals of, *115,* 117-118; Schwarzschild's concept, 104, 106; spinning (Kerr), 108; supermassive, possible locations of, 65, *94-95, 109,* 116-117; time line of theorists, *102-103;* and wormhole formation, 118, 124, 126; x-ray sources as, possible, *107,* 110, 116
Black Widow pulsar, 37, *48-49;* discovery of, 36-37
Bohr, Niels, 96, 97
Boötes void, 80, 82
Bothun, Gregory, 70-71, 72
Brecher, Kenneth: newspaper parody by, *105*
Brown dwarfs (celestial objects), 27-30; Forrest's search for, *28,* 29-30
Bubble chambers, 98, *99*
Bubble description of universe, 80, 82
Burbidge, Geoffrey, 67; quoted, 68

C
Campbell, Bruce, 29
Carbon clouds around R Coronae Borealis, theoretical, *46-47*
cD galaxies, 64
Centaurus A (galaxy): radio lobes from, *74*
Cepheid variables (stars), 43, 57; Delta Cephei, *42-43*
Cepheus (constellation): star-forming region in (Cepheus A), *38-39;* variable star in (Delta Cephei), *42-43*
Chandrasekhar, Subrahmanyan, 107
Chaos theory; chaotic systems, *17-21, 22-23;* Jupiter's Great Red Spot, *17, 18-19;* Pluto's orbit, 17, *20-21,* 23
Clark, Alvan and Alvan G., 31
Cline, Thomas, 15; quoted, 15-16
Cloud chambers, 98, *99*
Clusters of galaxies, 77-78, 79; Local Group, 77, 78, *86-87, 88,* 89; Perseus, *78-79;* Virgo, *75, 88,* 89. *See also* Superclusters
COBE (Cosmic Background Explorer; satellite), *118-119,* 120
Coleman, Sidney, 131
Coma cluster of galaxies, 79
Comets: Oort cloud of, *26*
Computer: Digital Orrery, *22,* 23
Computer simulations: of Jupiter's Great Red Spot, *18-19;* of Pluto's orbit, *20-21,* 23; in search for Planet X, 24
Cosmic Background Explorer *(COBE;* satellite), *118-119,* 120
Cosmic background radiation: COBE measurement of, *118,* 120; hot spot in, 87, 89
Cosmic jets, 65, *73-76*
Cosmic rays, 98
Cosmic strings, 93
Cowie, Lennox, 80
Cowley, Anne, 110
Crab, Southern (celestial object), *4-5*
Cronin, James, 101
Cygnus X-1 (x-ray source), *107,* 110

D
Dark matter, 77, 82, 93; missing mass explained by, 28, 72
Death Star (Nemesis; hypothetical body), *26-27*
Degeneracy pressure, 107
Delta Cephei (star), *42-43*
Desai, Upendra, 14
Digital Orrery (computer), *22,* 23
Dirac, Paul, 97; antiparticles predicted by, 97-98, 100
Dog Star (Sirius), *30,* 31-32
Doppler shift (redshift): of galaxies, study of, 58, 78-79, 80, 87; of quasars, 66-67, 68, 69
Dressler, Alan, 83
Dyson, Freeman: quoted, 16

E
Eclipsing star systems, 33-34, *35;* Black Widow pulsar, 37, *48-49*
Eggleton, Peter, 33
Einstein, Albert, 96, 97, *102;* and general theory of relativity, 103-104, 122; gravitational lensing predicted by, 70; and special theory of relativity, 103, 119, 129; and wormholes, 118
Einstein-Rosen bridges, 118. *See also* Wormholes
Electromagnetism and plasma filaments, 120-121
Electrons: and antimatter counterparts, 97-98, *99,* 100, *112;* in quantum theory, 96-97
Elliptical galaxies, 59, 60, *64;* with cosmic jets, *74-75*
Epsilon Aurigae star system, 33-34, *35*
Ergosphere, 108
Escape velocity, 101-102
Eta Carinae (star), 34-36, *36-37*
Event horizon, 104, *111*
Exotic-matter wormholes, *128-129*
Expansion of universe, 58, 77, *84-85*
Exploding galaxy (M82), *6-7*
Extinctions, mass, in Earth's history, 24, 26; Nemesis as cause of, *26*

F
Fitch, Val, 101
Ford, Joseph: quoted, 22
Forrest, William J., *28,* 29-30
Frame dragging, 108
Fruchter, Andrew, 36, 37

G
Galaxies, 52-93; active, *59,* 60, 63-65; associations of, 57, 77-78, *78-79, 81,* 82-83, *86-93;* catalogs of, 61, 62; center of Milky Way, radio images of, *12-13, 120;* classification of, 58-60, 64; COBE data's failure to explain, *118,* 120; controversies surrounding, 56; and cosmic jets, 65, 73, *74-76;* embryonic (protogalaxies), possible, 54-56, *55;* exploding (M82), *6-7;* failed, theory of, 70-72; history of study of, 56, 57-58; interactions of, 6, *52-53,* 60, *61-66;* peculiar motions of, 83, 85, *86-91;* population types of stars, Milky Way's, 39-40; and quasars, 67, 68, *69;* redshift, study of, 58, 78-79, 80, 87; with supermassive black holes, possible, 65, *94-95, 109,* 116-117; unseen matter in, 72, 77; voids between groups of, 79-80, *81,* 82
Gamma Arietis (star), *32*
Gamma rays, 98; bursts of, 14-16
Geller, Margaret, 80, 82
General theory of relativity: black holes and wormholes allowed by, 104, 106, 118, 124, 126; space-time in, 103-104, 106, *122-123*
Giclas 29-38 (star), 29
Giovanelli, Riccardo, 54, *55-*56
Gold, Thomas, 36
Gravitational lensing, 70
Gravity and gravity wells in space-time, 103-104, 106, *122-123. See also* Black holes
Great Attractor (supercluster of galaxies), 83, *91,* 93
Great Red Spot, Jupiter's, *17;* simulations of, *18-19*
Great Wall (supercluster of galaxies), *81,* 82
Gregory, Stephen, 79, 82
Gunn, James, 69

H
Haro, Guillermo, 38, 39
Harrington, Robert, 24
Hawking, Stephen, *103,* 108-109; theories about black holes, 109-110, 111, 112, 115; theory of multiple universes, 119, 131
Haynes, Martha, 54, 55, 56
HDE 226868 (star): and companion (Cyg X-1), *107,* 110; location of, *106*
Heisenberg, Werner, 97
Helium in Delta Cephei, 43
Herbig, George H., 38, 39

Herbig-Haro (HH) objects, *38-39, 73*
Hertzsprung, Ejnar, 30
HH111 (newborn star): jet of gas from, *73*
HH (Herbig-Haro) objects, *38-39, 73*
Homunculus nebula, 34, *36-37*
Hooke, Robert, 19
Hubble, Edwin, galactic studies of, 57-58, 78, 85; classification system, 58-60
Hubble constant, 58
Hubble flow: motions diverging from, 83, 85, 87
Huchra, John, 80, 82
Humason, Milton, 58, 78-79
Hydra-Centaurus supercluster of galaxies, 83, *89, 90,* 91
Hydrogen formations: galactic bridge, *62-63;* HH objects, *38-39;* protogalaxy, possible, 54, *55;* revealed by quasar spectra, 70
Hydroxyl masers of IRC + 10011, *44-45*

I
Impey, Chris, 71
Infrared and microwave radiation: *COBE* measurement of, *118;* from IRC + 10011, *44-45*
Infrared Astronomical Satellite (*IRAS*), 63
Infrared images of Herbig-Haro (HH) objects, *38-39*
Infrared search for brown dwarfs, *28,* 29-30
IRC + 10011 (celestial object), *44-45*
Irregular galaxies. *See* Peculiar galaxies
Irwin, Mike, 55

J
Jets, cosmic, 65, *73-76*
Jupiter (planet), Great Red Spot of, *17;* simulations of, *18-19*

K
Kerr, Roy, *103,* 108, 126
Kestevan 32 (supernova remnant), *76*
Kirshner, Robert, 80; quoted, 80, 82
Klebesadel, Ray, 14
Kumar, Sanjiv, 34

L
Lada, Charles: quoted, 39
Laplace, Pierre-Simon, *102*
Lapparent, Valerie de, 80
Large Magellanic Cloud (galaxy), 62; supernova remnant in, 15; x-ray source in, 110
Leavitt, Henrietta, 57
Lensing, gravitational, 70
Light, gravity's effect on, 122. *See also* Black holes

Light curves, 41; of Black Widow pulsar, *49;* of Delta Cephei, *43;* of IRC + 10011, *45;* of Nova Cygni 1975, *51;* of R Coronae Borealis, *47*
LINER (low-ionization nuclear-emission region) galaxies, 64-65
LkCa 4 (star): infrared image, *28*
LMC X-3 (x-ray source), 110
Local Group of galaxies, 77, 78, *86-87, 88,* 89
Local Supercluster of galaxies, 83, 89, *90, 92*
Low-surface-brightness galaxies, theory of, 71-72
Lynden-Bell, Donald, 116

M
M81 (galaxy), 6
M82 (galaxy), *6-7*
M87 (galaxy): jet of gas from, *75*
M106 (galaxy), *59*
McMahon, Richard, 55
Maffei, Paolo: quoted, 34
Magentic field of Nova Cygni 1975, 51
Malin, David, 71
Malin 1 (galaxy), 71-72
Marcus, Philip: computer simulation by, *18-19*
Markarian galaxies, 64
Marsden, Brian: quoted, 24
Masers, hydroxyl, of IRC + 10011, *44-45*
Mass transfer theory of binary stars, 31-32, *33*
Meteorites, 23; and mass extinctions, 24, 26, *27*
Michell, John, 101-102
Microwave and infrared radiation: *COBE* measurement of, *118;* from IRC + 10011, *44-45*
Milky Way (galaxy), 62; center, radio images of, *12-13, 120;* in expanding universe, *84-85;* populations of stars in, 39-40
Millisecond pulsars, 36; Black Widow as, 37, *48-49*
Mini (primordial) black holes, 110, 111, *114,* 115; explosion of, 110, *115*
Missing mass in universe, 28, 72
Moons of Neptune: theory about, *25*
Morris, Michael, 119, 129

N
N49 (supernova remnant), 15
Nebulae: around Eta Carinae (Homunculus), 34, *36-37;* galaxies seen as, 56, *57;* around R Aquarii, *32-33;* and star births, *27*
Nemesis (hypothetical body), 26-27
Neptune (planet): moons, theory about, *25;* orbital irregularities of, 24

Neutron stars, 15, 107; pulsars, 36-37, *48-49*
N galaxies, 64
NGC 1097 (galaxy): jets from, apparent, *76*
NGC 5754 (galaxy), *52-53*
NGC 6166 (galaxy), *66*
Nordstrom, Gunnar, 107, 126
Nova Cygni 1975 (star), *50-51*
Nova formation, 51

O
Oort, Jan, 72
Oort cloud, *26*
Oppenheimer, J. Robert, *102,* 107-108
Orbits, effects on: of asteroids, 23; and Planet X theory, 24, *25;* of Pluto, 17, *20-21,* 23
Orrery, *22;* Digital Orrery (computer), *22,* 23
Osmer, Patrick, 68
Ostriker, Jeremiah, 80

P
Particles and antiparticles, 97-98, *99,* 100-101; virtual, and black holes, 109, *112-113*
PC 1158 + 4635 (quasar), 68-69
Peculiar galaxies, *62-66;* the Mice, *65;* NGC 6166, *66;* ring shaped, *64;* Toadstool, *62-63*
Peculiar motions of galaxies, 83, 85, *86-91*
Penrose, Roger, *103,* 108
Peratt, Anthony, 121
Perseus cluster of galaxies, *78-79*
Perseus-Pisces-Pegasus supercluster of galaxies, *79,* 92
Planets: behavioral anomalies of, *17-21,* 23-24, *25;* formation of, 27
Planet X (hypothetical body), 24, *25*
Plasma filaments: in Milky Way, *120;* and origins of universe, 120, 121
Pluto (planet), 23-24; chaos in orbit of, 17, *20-21,* 23
Population I, II, and III stars, 39-40
Positrons, 98, *99,* 100; in virtual pairs, *112*
Primordial (mini) black holes, 110, 111, *114,* 115; explosion of, 110, *115*
Pringle, James, 33
Protogalaxies, possible, 54-56, *55*
Protostars, 27, 73
Pulsars, 36-37, *48-49*
Pup (Sirius B; star), 30, 31, 32
Puppis A (supernova remnant), *2-3*

Q
Quantum theory, 96-97; exotic matter, 129; fluctuations, 119, 131;

141

Hawking's application of, to black holes, 109, 111, 112
Quasars, 65-70, *69;* binary system, *109;* black holes in, theory of, 109, 116-117; first discoverer of (Maarten Schmidt), 66, 68, *71;* jet from, *75;* as white holes, 117

R

Radio astronomy, studies in, 65-66; Andromeda, map of, *94-95;* galactic pairs, enhanced emissions from, 62-63; lobes from cosmic jets, 65, 73, *74;* Milky Way's center, *12-13, 120;* protogalaxy, possible, discovery of, 54, *55;* pulsar emissions, 36, *49;* quasar emissions, 66, *75;* R Aquarii, *30-31;* supernova remnants, *2, 76*
R Aquarii (star), 32-33; in radio image, *30-31*
R Coronae Borealis (star), *46-47*
RCrB variables (stars), 47
Redshift: of galaxies, study of, 58, 78-79, 80, 87; of quasars, 66-67, 68, 69
Red spot, Jupiter's, *17;* simulation of, *18-19*
Reissner, Heinrich, 106-107, 126
Reissner-Nordstrom singularity, 107
Reissner-Nordstrom wormholes, 126
Relativity, theory of: black holes and wormholes allowed by, 104, 106, 118, 124, 126; space-time in, 103-104, 106, *122-123;* special, 103, 119, 129; and white holes, 117, 118
Rosen, Nathan, 118
Rubin, Vera, 72, 77
Russell, Henry Norris, 30
Ryle, Martin, 65

S

Sagan, Carl, 118-119
Satellites: *COBE, 118-119,* 120; *IRAS,* 63; Vela, 14
Schmidt, Maarten, 66, 68, 69, *71*
Schneider, Donald: quoted, 68-69
Schwarzschild, Karl, *102,* 104, 106, 126
Schwarzschild radius, 104, *111*
Schwarzschild wormholes, *126-127*
Seidelmann, Kenneth, 24
Seven Samurai (team of scientists), 83
Seyfert, Carl, 60
Seyfert galaxies, 60, 64
Silicon concentrations in Gamma Arietis, *32*
Singularities, *111,* 124; Reissner-Nordstrom, 107; ring shaped (Kerr), 108; Schwarzschild, 104, 106; and space-time, 118, *124-125*
Sirius A and B (stars), *30,* 31-32

Southern Crab (celestial object), *4-5*
Space-time: in general relativity, 103-104, 106, *122-123;* and quantum fluctuations, 119, 131; singularities penetrating, in wormhole formation, 118, *124-125*
Special theory of relativity, 103; and time travel through wormhole, 119, 129
Spiral galaxies, 58-60; Andromeda, 57-58, *94-95;* M81, 6; M106, *59;* NGC 1097, apparent jets from, *76;* NGC 5754, *52-53;* and quasar, *69;* rotation rates of, 72, 77
Starburst galaxies, 63
Stars: birth of, 27, 34, 63; black holes from, 107-108, 124; with brown dwarfs as companions, possible, *28-*29; cosmic jets from, *73;* Death Star (Nemesis), hypothetical, *26-*27; Epsilon Aurigae, 33-34, *35;* Eta Carinae, 34-36, *36-37;* evolution of, 30-31; Gamma Arietis, *32;* and HH objects, 38-39, *73;* neutron, 15, 36-37, *48-49,* 107; Population I, II, and III, 39-40; Sirius A and B, *30,* 31-32; T Tauri stage of, 37-38; with x-ray sources as companions, *106-107,* 110. *See also* Galaxies; Variable stars
Superclusters of galaxies, 79, 83; Great Attractor, 83, *91,* 93; Great Wall, *81,* 82; Hydra-Centaurus, 83, *89, 90,* 91; Perseus-Pisces-Pegasus, 79, *92*
Supermassive black holes, possible locations of, 65, 116-117; Andromeda, *94-95;* binary quasar system, *109*
Supernova remnants: Kestevan 32, *76;* N49, 15; neutron stars as, 15, 36, 49; Puppis A, *2-3;* Zwicky's search for, 60-61
Sussman, Gerald: Pluto's orbit studied by, *20-21,* 23
Synchrotron radiation in Andromeda, *94-95*

T

Taurus (constellation): brown dwarfs in, possible, *28*
Telescopes: infrared, 63; 100-inch, Mount Wilson Observatory's, 57; radio, largest, 36, 54
Theory of relativity. *See* Relativity, theory of
Thompson, Laird, 79
Thorne, Kip, *116,* 119, 129
3C273 (quasar), 66-67, *71;* jet from, *75*
Tifft, William, 79
Time: black hole's effect on, 106; reversal of, and white holes, 117; travel in, through wormhole, 119, 129
Toadstool (galaxies), *62-63*
Tombaugh, Clyde, 23-24
Transfer theory of binary stars, 31-32, 33
T Tauri winds, 37-38
Type 1 and Type 2 Seyfert galaxies, 64

U

Universe: expansion of, 58, 77, *84-85;* ultimate fate of, question of, 77, 93. *See also* Big Bang
Universes, multiple: wormholes and, 119-120, *130-131*
Uranus (planet): orbital irregularities, 24

V

Van den Heuvel, Edward, 31
Variable stars, 41, *42-51;* Cepheids, *42-43,* 57; IRC + 10011, *44-45;* Nova Cygni 1975, *50-51;* pulsar, 36-37, *48-49;* R Aquarii, *30-31,* 32-33; R Coronae Borealis, *46-47*
Vaucouleurs, Gérard de, 78
Vela satellites, 14
Virgo cluster of galaxies, *75, 88,* 89
Virtual particles and black holes, 109, *112-113*
Vorontsov-Velyaminov, Boris, 61-62
Voyager 1 photo of Jupiter, *17*

W

Walborn, Nolan, 35
Weak nuclear force, 101
Weedman, Daniel: quoted, 69-70
Wheeler, John, 104, 110, *116,* 118, 119
White holes, 117-118; exploding primordial black holes as, *115*
Winds: T Tauri, 37-38
Wisdom, Jack, *23;* Pluto's orbit studied by, *20-21,* 23
Wolfe, Arthur: quoted, 55
Wormholes, 116, 118-119, *124-125;* Kerr, 126; and multiple universes, 119-120, *130-131;* Reissner-Nordstrom, 126; Schwarzschild, *126-127;* traversable, 119, *128-129*

X

X-ray sources as black holes, possible, *107,* 110, 116

Y

Yorke, James, 22

Z

Zuckerman, Benjamin, 29
Zwicky, Fritz, *60-61,* 72, 77-78, 107

ACKNOWLEDGMENTS

The editors wish to thank Halton Arp, Max Planck Institut für Physik und Astrophysik, Garching, Germany; Jack O. Burns, New Mexico State University, Las Cruces; Thomas L. Cline, NASA Goddard Space Flight Center, Greenbelt, Md.; Kris Davidson, University of Minnesota, Minneapolis; Robert S. Harrington, U.S. Naval Observatory, Washington, D.C.; Charles J. Lada, University of Arizona, Tucson; Adair Lane, Boston University, Boston; Jean J. Lorre, Jet Propulsion Laboratory, Pasadena, Calif.; Julie H. Lutz, Washington State University, Pullman; Andrew Michalitsianos and Malcolm Niedner, NASA Goddard Space Flight Center, Greenbelt, Md.; Sten Odenwald, Naval Research Laboratory, Washington, D.C.; Frazer Owen, National Radio Astronomy Observatory, Socorro, N.Mex.; Wolfgang Priester, Bonn, Germany; James Pringle, Johns Hopkins University, Baltimore; Axel Quetsch, Max Planck Institut für Astronomie, Heidelberg, Germany; Tom Rickey, University of Rochester, Rochester, N.Y.; Robert S. Roger, Herzberg Institute of Astrophysics, Dominion Radio Astrophysics Observatory, Penticton, British Columbia; Lawrence Rudnick, University of Minnesota, Minneapolis; Hugo E. Schwarz, European Southern Observatory, Santiago, Chile; Patricia Smiley, National Radio Astronomy Observatory, Charlottesville, Va.; Peter Stockman, Space Telescope Science Institute, Baltimore; Jack W. Sulentic, University of Alabama, Tuscaloosa; Harry Swinney, University of Texas, Austin; Saul Teukolsky, Cornell University, Ithaca, N.Y.; Steven Vogt, University of California, Santa Cruz; Richard M. West, European Southern Observatory, Garching, Germany; Gerd Wielebinski, Max Planck Institut für Radioastronomie, Bonn, Germany; Lee Anne Willson, Iowa State University, Ames; Andrew S. Wilson, University of Maryland, College Park; and P. Frank Winkler, Middlebury College, Middlebury, Vt.

PICTURE CREDITS

The sources for the illustrations in this book are listed below. Credits from left to right are separated by semicolons; credits from top to bottom are separated by dashes.
Cover: Art by Matt McMullen. Front and back endpapers: Art by Time-Life Books. 2, 3: Radio data from D. K. Milne et al., optical plate taken with the UK Schmidt telescope; courtesy National Radio Astronomy Observatory, Charlottesville, Va. 4, 5: Hugo E. Schwarz, European Southern Observatory. 6, 7: Science Photo Library/Photo Researchers. 12, 13: Courtesy NRAO/AUI. 14: Initial cap, detail from pages 12, 13. 17: NASA/Jet Propulsion Laboratory. 18, 19: Philip Marcus, Nicholas Socci, University of California, Berkeley (5); Dr. Harry Swinney, University of Texas, Austin. Line art by Time-Life Books. 20, 21: Art by Damon M. Hertig. 22: Courtesy Collection of Historic Scientific Instruments, Harvard University; Gerald Sussman, Massachusetts Institute of Technology. 23: Jack Wisdom, Massachusetts Institute of Technology. 25: Line art by Time-Life Books—art by Fred Devita. 26: Art by Fred Devita. 28: William Forrest, University of Rochester; photograph by James Montanus, University of Rochester. 30, 31: Lick Observatory, University of California, Santa Cruz—NASA/ESA. 32, 33: Lick Observatory, A. Hatzes, G. Penrod, S. Vogt. 35: Art by Fred Devita. 36, 37: Courtesy K. Davidson, University of Minnesota, image made at Cerro Tololo Inter-American Observatory. 38, 39: Courtesy Adair P. Lane, Boston University and John Bally, AT & T Bell Laboratories. 42-51: Art by Rob Wood of Stansbury, Ronsaville, Wood, Inc., except graphs by Fred Holz. 52, 53: Dr. William C. Keel/Science Photo Library. 54: Initial cap, detail from pages 52, 53. 55: Line art by Time-Life Books based on radio map provided by Riccardo Giovanelli and Martha Haynes, National Astronomy and Ionosphere Center, Arecibo Observatory. 59: Copyright 1988 by Dr. James D. Wray. 61: Fritz Zwicky, "Multiple Galaxies," in *Astrophysics IV: Stellar Systems*, Vol. 53 of *Encyclopedia of Physics*, edited by S. Flugge, Berlin-Göttingen-Heidelberg, Springer-Verlag, 1959; California Institute of Technology. 62, 63: National Optical Astronomy Observatories, Tuscon, Ariz. 64, 65: Marshall Joy of NASA, Victor Blanco of Cerro Tololo Inter-American Observatory, James Higdon of University of Texas, Austin, and NASA; NOAO. 66: Rudolph Schild and Thomas Stephenson, Harvard-Smithsonian Center for Astrophysics. 69: Halton Arp, Max Planck Institut für Physik und Astrophysik, Garching, Germany; courtesy David A. Pierce. 71: California Institute of Technology. 73: Bo Reipurth, European Southern Observatory. 74, 75: Background photo by David Malin/Anglo-Australian Observatory. Line art by Time-Life Books based on radio map by David Clarke and Jack Burns. 75: Jean Lorre, JPL/NASA—Jodrell Bank, London. 76: Jean Lorre/Science Photo Library; courtesy R. S. Roger, D. K. Milne, M. J. Kesteven, K. J. Wellington, R. F. Haynes. 78, 79: NOAO; line art based on graph provided by John Huchra, Harvard Smithsonian Center for Astrophysics. 81: Margaret J. Geller, John P. Huchra, Emilio E. Falco, Robert McMahan, Harvard Smithsonian Center for Astrophysics. 84-93: Art by Stephen R. Wagner. 94, 95: R. Beck, E. Berkhuigsen, R. Wielebinski, Max Planck Institute für Radioastronomie, Bonn. 96: Initial cap, detail from pages 94, 95. 99: Lawrence Berkeley Laboratory, University of California, Berkeley. 102, 103: Courtesy the Royal Society, London; Hulton-Deutsch Collection, London; Yerkes Observatory, University of Chicago, Williams Bay, Wis.; AIP Neils Bohr Library, Physics Today Collection; courtesy Roy Kerr; Anthony Howarth, Woodfin Camp; Camera Press Limited, London. Background art by Time-Life Books. 105: Courtesy Professor Kenneth Brecher, Boston University. Background photograph by Dennis di Cicco/*Sky & Telescope* magazine. 106, 107: Dennis di Cicco/ *Sky & Telescope* magazine; F. Seward of NASA, C. Jones and C. Stein of Smithsonian Astrophysical Observatory. 109: G. Meylan of Space Telescope Science Institute, G. Djorgovski of California Institute of Technology, P. Shaver of European Southern Observatory. 111-115: Art by Stephen R. Bauer. 116: Courtesy Robert P. Matthews, Joseph Henry Laboratory, Princeton University; courtesy David A. Pierce. 118, 119: Line art by Time-Life Books based on chart provided by NASA; NASA. 120, 121: NRAO/AUI, F. Yusef-Zadeh, M. R. Morris, D. R. Chance. 122-131: Art by Matt McMullen.

Time-Life Books is a division of Time Life Inc., a wholly owned subsidiary of
THE TIME INC. BOOK COMPANY

TIME-LIFE BOOKS

PRESIDENT: Mary N. Davis

Managing Editor: Thomas H. Flaherty
Director of Editorial Resources:
Elise D. Ritter-Clough
Director of Photography and Research:
John Conrad Weiser
Editorial Board: Dale M. Brown, Roberta Conlan, Laura Foreman, Lee Hassig, Jim Hicks, Blaine Marshall, Rita Thievon Mullin, Henry Woodhead
Assistant Director of Editorial Resources/ Training Manager: Norma E. Shaw

PUBLISHER: Robert H. Smith

Associate Publisher: Trevor Lunn
Editorial Director: Donia Steele
Marketing Director: Regina Hall
Production Manager: Marlene Zack
Supervisor of Quality Control: James King

Editorial Operations
Production: Celia Beattie
Library: Louise D. Forstall
Computer Composition: Deborah G. Tait (Manager), Monika D. Thayer, Janet Barnes Syring, Lillian Daniels
Interactive Media Specialist: Patti H. Cass

Correspondents: Elisabeth Kraemer-Singh (Bonn), Christine Hinze (London), Christina Lieberman (New York), Maria Vincenza Aloisi (Paris), Ann Natanson (Rome).

VOYAGE THROUGH THE UNIVERSE

SERIES DIRECTOR: Roberta Conlan
Series Administrator: Susan Stuck

Editorial Staff for *Cosmic Mysteries*
Designer: Robert Herndon
Associate Editor: Kristin Baker Hanneman (pictures)
Text Editor: Robert M. S. Somerville
Researchers: Mark Galan, Karin Kinney, Mary H. McCarthy
Writers: Stephanie Lewis, Barbara Mallen
Assistant Designer: Barbara M. Sheppard
Copy Coordinator: Darcie Conner Johnston
Picture Coordinator: Jennifer Iker
Editorial Assistant: Katie Mahaffey

Special Contributors: James Cornell, James Dawson, Michael Lemonick, Peter Pocock, Chuck Smith, Mark Washburn (text); Edward Dixon, Dan Kulpinski, Todd Lang, Jocelyn G. Lindsay, Jacqueline L. Shaffer, Roberta Yared (research); Barbara L. Klein (index).

CONSULTANTS

JOHN M. BLONDIN is a research associate at the University of Virginia in Charlottesville, where he investigates astrophysical gas dynamics.

ALAN H. BRIDLE, an astronomer at the National Radio Astronomy Observatory in Charlottesville, is an observer of cosmic jets.

SIDNEY COLEMAN is a theoretical physicist and professor of physics at Harvard University.

ALAN M. DRESSLER is an astronomer at the Carnegie Institution of Washington's Mount Wilson and Las Campanas Observatories. He explores the formation and evolution of galaxies.

STEPHEN A. GREGORY is an astronomer affiliated with the University of New Mexico's Department of Physics and Astronomy. His area of interest is galaxy superclusters.

JOHN P. HUCHRA, an observational cosmologist, is a professor of astronomy at Harvard University and the associate director of the Harvard Smithsonian Center for Astrophysics.

JAMES B. KALER, an expert on spectroscopy, teaches stellar astronomy at the University of Illinois.

ALAN P. LIGHTMAN is a professor of physics at Massachusetts Institute of Technology. His special research interests are relativity theory and high-energy astrophysics.

STEPHEN MARAN, a senior staff scientist at NASA Goddard Space Flight Center, Greenbelt, Maryland, writes widely in astronomy and is the press officer for the American Astronomical Society.

PHILIP MARCUS is a professor with the Department of Mechanical Engineering at the University of California, Berkeley, where he researches fluid dynamics in astrophysics, the transition to chaos, and computational physics.

MICHAEL MORRIS is a relativity theorist in the Department of Physics at the University of Wisconsin, Milwaukee, where he studies space-time curvature and quantum cosmology.

DON NELSON PAGE is a professor of physics at the Pennsylvania State University and a visiting professor of physics at the University of Alberta. He researches black holes and the early universe.

JACK WISDOM is a physicist at the Massachusetts Institute of Technology, where he specializes in chaos theory.

Library of Congress Cataloging in Publication Data
Cosmic mysteries/by the editors of Time-Life Books.
  p. cm. — (Voyage through the universe)
Includes bibliographical references and index.
ISBN 0-8094-9062-5 (trade)
ISBN 0-8094-9063-3 (lib. bdg.)
1. Astronomy. 2. Astrophysics.
I. Time-Life Books.
II. Series.
QB43.2.C68 1992
520—dc20   91-45193
                                           CIP

For information on and a full description of any of the Time-Life Books series, please call 1-800-621-7026 or write:
Reader Information
Time-Life Customer Service
P.O. Box C-32068
Richmond, Virginia 23261-2068

© 1990 Time-Life Books Inc. All rights reserved. No part of this book may be reproduced in any form or by any electronic or mechanical means, including information storage and retrieval devices or systems, without prior written permission from the publisher, except that brief passages may be quoted for reviews.
Revised edition. First printing 1992. Printed in U.S.A.
Published simultaneously in Canada.
School and library distribution by Silver Burdett Company, Morristown, New Jersey 07960.

TIME-LIFE is a trademark of Time Warner Inc. U.S.A.

REVISIONS STAFF

EDITOR: Roberta Conlan

Associate Editor/Research: Quentin G. Story
Art Director: Robert K. Herndon
Assistant Art Director: Kathleen Mallow
Picture Coordinator: David Beard

Consultant: Stephen P. Maran. *See consultants.*